# Black box optimization with exact subsolvers

## subsolvers

### A radial basis function algorithm for problems with convex constraints

**Dissertation**

zur Erlangung des akademischen Grades eines

Doktors der Naturwissenschaften

(Dr. rer. nat.)

Dem Fachbereich IV der Universität Trier

vorgelegt von

**Christine Edman**

Trier, 2016

Eingereicht am 10.05.2016

Gutachter: Prof. Dr. Mirjam Dür
          Prof. Dr. Ralf Werner

Tag der mündlichen Prüfung: 07.07.2016

Bibliografische Information der Deutschen Nationalbibliothek

Die Deutsche Nationalbibliothek verzeichnet diese Publikation in der
Deutschen Nationalbibliografie; detaillierte bibliografische Daten sind
im Internet über http://dnb.d-nb.de abrufbar.

ISBN 978-3-8325-4329-7

Logos Verlag Berlin GmbH
Comeniushof, Gubener Str. 47,
10243 Berlin
Tel.: +49 (0)30 42 85 10 90
Fax: +49 (0)30 42 85 10 92
INTERNET: http://www.logos-verlag.de

# Contents

# Chapter 1

# Introduction

## 1.1 Expensive optimization problems

The subject of this thesis is the treatment of expensive optimization problems. These are mathematically formulated problems where the objective function to be minimized is not available in analytical form. Therefore derivatives or other analytical instruments cannot be used. Moreover, every evaluation of the objective function will be expensive in some way. Objective functions can be expensive due to computing time, consumption of resources or costs.

As an example, it is often impossible to predict the reaction of chemicals. However, it is desirable to find the ideal composition of the individual substances to produce a material as tear-resistant as possible for example. If a chemical experiment causes a lot of effort, then it is worthwhile to reduce the number of experiments. Other examples are soil contamination and the question where the highest concentration of contamination is. Although single measurements can be made, there is no analytical representation of the function to be optimized. As a result, an assessment of the quality of these single measurements is therefore not possible. Obviously starting a series of arbitrary evaluations is useless. There are numerous examples of those inherent optimization problems.

More formally, we will consider the global optimization problem

$$\min f(x) \quad \text{s.t. } x \in R. \tag{1.1}$$

Here $f : \mathbb{R}^d \to \mathbb{R}$ is the expensive objective function and $R \subset \mathbb{R}^d$ is in general a convex and compact set that contains all possible input combinations. For our approach $R$ will be a $d$-dimensional rectangle. For simplicity we will assume that

$R$ is the unit rectangle $[0,1]^d$. This is no loss of generality due to the fact that every box can be transformed to a unit box by an invertible map. Moreover we will understand the "expensiveness" of $f$ as a lack of analytical representation, so that every evaluation causes enormous cost in terms of money, computation time or the use of other resources. Because of this lack of analytical representation these functions are also called "black box functions". The optimal parameters to be found, like the single chemicals, ranges in these examples within an interval which is bounded by a minimal and a maximal value. The condition in (1.1) is in this case simply a $d$-dimensional rectangle, where the dimension $d$ is the number of parameters to be computed. See also [42, 27, 5] for possible real-world applications.

## 1.2 Global optimization methods

Global optimization methods have been investigated intensively over the last few decades. Techniques have become more and more developed and finetuned. As a result, the different optimization methods cannot always be distinguished sharply from each other. But they can at least roughly be catalogized into the three groups stochastic methods, deterministic methods and response surface methods, where the latter will be treated in a separate section.

### 1.2.1 Stochastic and deterministic methods

Stochastic methods, like multistart algorithms or random search, usually incorporate a stochastic element within the search strategy. Within multistart algorithms, for example, several starting points are randomly generated and then local searches are started. Other stochastic methods imitate or simulate the behaviour, or phenomena, of nature by using stochastic tools. This group of optimization techniques include simulated annealing, evolutionary algorithms and algorithms that use the concept of swarm intelligence. While simulated annealing simulates a physical phenomen, that is to say a cooling process of smelted materials, the other two concepts imitate the behaviour of animals or the evolution in general. Stochastic methods theoretically find the global optimum with probability one, but the number of iterations may increase rapidly until the algorithm even comes near an optimum. The previously mentioned stochastic methods mostly act independently of the objective function. Methods that use statistical models and increasingly include information of the objective function, were, amongst other things, influenced by the approximation theory in the beginning of the 1980s. Initiated by approximation methods for one-dimensional functions [35] the multi-dimensional optimization theory, using statistical models, was extended [44] and led to to the well-known Baysian methods, which are no longer heuristics and belong to the response surfaces, which will be discussed in Section 1.2.2.1.

Deterministic methods usually use structual information about the objective function. This may be a parameter like the Lipschitz constant [31, 16], convexity, or the information that the objective function is a d.c. function, i.e. it can be written as the difference of two convex functions. As the aim is to find the global optimum, it is obvious to use such a global information. The main drawback of this use is that, even if the function is Lipschitz, it is not guaranteed that the Lipschitz constant can be easily computed. Likewise, a d.c. decomposition of the objective may be hard to find. Typically deterministic algorithms use these parameters for computing lower or upper bounds within a Branch and Bound routine. Also convex envelopes are widely used to compute lower bounds [29, 12]. One method, for example, uses a convex underestimator which needs the computation of parameters $\alpha_i$. These parameters are directly linked to properties of the Hessian matrix, and the algorithms incorporating these parameters are called $\alpha$BB [4, 1, 3, 2, 32]. The computation of these parameters takes a lot of effort, uses global information and can only be used for $C^2$-continuous functions. A deterministic algorithm that does not use any parameters is the Lipschitzian optimization algorithm method "DIRECT" [25], which paradoxically does not need a specific Lipschitzian constant. This algorithm was sucessful in practical applications and involves a strategy that identifies a set of so called "potential intervals" for divisions by considering "all possible Lipschitz constants" simultaneously. Global information is not used. Instead, the search area is explored thoroughly. Therefore, this method still requires many sample points. Good overviews about stochastical and deterministic optimization strategies are given in [9, 21, 11].

## 1.2.2 Response surfaces

Response surfaces are model functions that rebuild the original object function more or less well. Some of the methods discussed in Section 1.2.1 contain such a model. The statistical model of course works as a response surface and the convex underestimator for the $\alpha$BB-algorithm can be seen as a model for the objective as well, even though its purpose is the computation of lower bounds. In contrast, the DIRECT-algorithm and the heuristical stochastic methods in Section 1.2.1 do not use any model function. In order to solve optimization problems with an expensive objective, response surfaces are very useful due to the fact, that its function evaluations are usually very cheap. In addition, if the response surface rebuilds the expensive objective well, the evaluations can give some valuable information for the optimization process. The benefit of using model functions in this context is obvious. Instead of evaluating the (expensive) objective function too many times, the model function can be used to decide where to put further sample points. Optimization methods based on response surfaces can be roughly classified with respect to two aspects: first, with respect to the type of response surface which is used, and second, with respect to the method used to select the next evaluation point [24]. A further aspect, which is essential for the success of a global optimization method, is the interaction between these two aspects, see [24].

### 1.2.2.1 Types of response surfaces

Response surfaces can either be interpolating surfaces or approximate the given sample points according to some distance measure. For simplicity we consider only interpolating surfaces. This means that for given sample points $x^1, \ldots, x^n \in R$ a model, the response surface, is constructed which matches the points $(x^1, f(x^1)), \ldots, (x^n, f(x^n))$, where $f$ is the original objective function. The so called "interpolating model" can be constructed in many ways: it can be a quadratic function (only for a sufficiently small set of sample points), the sum of weighted basis functions with static parameters (e.g. "radial basis functions", see the following sections), the sum of weighted basis functions with optimized parameters ("kriging") or it can be the conditional mean function of a conditioned stochastic process. Those last methods belong to the group of "Baysian methods". Baysian methods, like e.g. the "Expected improvement", use a stochastic model and a utility function for the optimization process. In general the Bayesian methods differ in the utiliy function which is used to select the next sample point. They all have in common that this utility function only depends on the sample points already made. A further algorithm is the "EGO-Algorithm" [26] which incorporates a principle similar to the Bayes methods. Here a (simple) regression model is used for the response surface instead of a stochastic model. The main difference to Bayes methods is that in every step of the algorithm the estimations of the parameters for the response surface are updated corresponding to the maximum likelihood of the sampled points.

### 1.2.2.2 Principles of selecting the next evaluation point

We have seen in Section 1.2.2.1 that there exist many ways to build response surfaces for given sample points. A naive way of chosing the next sample point would be to just minimize the response surface and use the minimum point for the next expensive evaluation, build the next response surface and so on. This strategy does not always lead to a local minimum of the original function [24].

Within Baysian methods and the EGO-Algorithm different utility functions are optimized to compute a next sample point. Nevertheless, both sorts of methods have in common that the utility function whose maximum is selected to be the next evaluation point, is based on parameters that were calculated from the observed data only. The search strategies may look sophisticated - but the parameters as well as the stochastic models are built exclusively from the already observed data. No further input that maybe improves the search strategy is involved. Therefore these parameters depend strongly on the initial sample and can, in the worse case, be very misleading for the optimization process (see also [24]).

With the so called "P-Algorithm" [28] a different idea of search strategy within a Bayesian method was introduced. The name "P-algorithm" is due to the fact that maximizing the utility function corresponds to maximizing a specific probability. The innovative idea is to maximize the probability of achieving a certain target function value $F_n^* - \epsilon_n$ below the global minimum value $F_n^*$ of the considered utility function. The use and necessary exact specification of the parameter $\epsilon_n$ indicates, that here an additional information, or "hypothetical assumption", is used which does not depend on the sampled points alone. The "hypothetical assumption" here is that the target function value $F_n^* - \epsilon_n$ is close (or even equal) to the real optimal minimum value of the original objective.

Algorithms with an explicit formulation of a "hypothetical assumption" were afterwards proposed without using a stochastical model. The point of origin for developing these algorithms is a technique proposed [13] by D.R. Jones. He uses the well known minimal property of natural cubic splines to employ the innovative idea of choosing a target value. Natural cubic splines are used for interpolation and are typically computed by solving a linear system of equations. Now the integral of the natural cubic splines serves as a "measure of bumpiness" and a special auxiliary function is used to compute the next sample point. With this connection to the research field of approximation theory, Jones was able to add a variable interpolation condition (hypothetical assumption) to the system of equations for calculating the response surface. Algorithms that followed that idea are for example the "Radial basis function methods (RBF methods)" published by Gutmann [13] in 2001, which we will focus on in the following Chapter 2.

# Chapter 2

# Optimization methods for expensive problems

## 2.1 Surface methods

In the case of expensive problems the aim is to use as few function evaluations as possible and at the same time to approximate a global minimum. Therefore it is very important to develop a sophisticated strategy for determining the evaluation points. The main and simple idea of surface methods is to build a model of the original objective function and then to decide, depending on this model, where to put the next sample point. To get this model, usually a few sample points of the original expensive objective function, also called "black box function", are needed. The model, which matches the sample points, consists of analytically tractable functions and is cheap to evaluate. With this model we decide where to sample the expensive function next. After sampling the next point, a new model, which matches all already sampled points, is built and the procedure continues. Of course the success of applying the model depends on how accurately it approximates the original function. On the other hand, the shape of a black box function is never known, which makes the comparison of the model and the objective function difficult. It is nearly impossible to assess the accuracy of the model. Furthermore it is important to use the scarce pieces of information about the black box function as efficiently as possible. So the surface methods differ not only in the choice of model functions but also mainly in the approach of searching for the next point to evaluate the black box function. For a detailed classification of different surface methods or response surface approaches see [24].

## 2.2 Surface methods with radial basis functions

In Section 2.1 the necessary use of analytically tractable functions within the surface model is emphasised. One possible option are the so-called *radial basis functions*. In recent years the interest in using radial basis functions within surface methods has grown rapidly. Originally this movement was based on the research of the special radial basis functions *thin plate splines* and *multiquadrics* in the 1970s. The thin plate splines were studied by Duchon in 1970 (see [37]) and the multiquadrics by Hardy in 1971 [17]. The authors figured out some remarkable characteristics concerning the approximation and interpolating behaviour of these functions. Up to now there exist four types of radial basis functions, which will be listed in Section 2.3.1.1, Definition 2.1. Some authors developed efficient algorithms (see [13], [38]) involving radial basis functions. Radial basis functions were used for interpolating models (see [38]) as well as for approximation models (see [22]) within optimization methods, within trust-region-methods [46], and for solving mixed-integer-problems [18].

A very important factor for every surface method is the strategy of choosing the next evaluation point when a model of the original expensive function is already constructed. In this point the authors differ and many different strategies have been developed. A sophisticated choice of the evaluation points is necessary since the expensive objective function does not allow many trials. To point out the main factors which influence the quality of the choice of the evaluation points, we will summarize the auxiliary tasks within surface methods: There are mainly two tasks that have to be done. First, to create an objective function as similar as possible to the real expensive function and second to find a good search strategy for evaluation points that leads as close as possible to the optimal solution. In the given examples in Section 1.1 the best of the evaluated sample points would then finally be used to produce the chemical composition or to vanish the soil contamination. Obviously, the success of solving the second task depends on how accurately the first task has been solved. The better the model function represents the original objective function, the easier a promising sample point can be found. On the other hand, creating a function that has to represent an unknown function works much better when characteristic points like minima are known and considered within the optimization process. Therefore both problems affect each other and it is recommended to treat them simultaneously. In his thesis [13] Gutmann considers this mutual influence and develops an appropriate algorithm: the radial basis function method (RBF method). This means that the search strategy for the next point to be evaluated not only relies on the model developed so far, but also on an assumption (or hypothesis) about the (unknown) optimum of the objective function. In Section 2.3 we will explain how the RBF method works and what are the main inherent problems that have to be solved.

## 2.3 Gutmann's RBF method

In this section we introduce the RBF algorithm that was developed by Gutmann [14] in 2001. Recall that we consider the global optimization problem

$$\min f(x) \quad \text{s.t. } x \in R.$$

with the expensive function $f : \mathbb{R}^d \to \mathbb{R}$ and the convex and compact set $R \subset \mathbb{R}^d$. We continue with a very short pseudo code of the RBF method, Algorithm 2.1, to emphasize the inherent problems that have to be solved.

Up to now we have not discussed how we compute the "smoothest function" in Algorithm 2.1, Step 1., what "smoothness" means mainly in the multidimensional case and how the problems in Steps 2.a) and b), can be solved. We will see in the next sections that we need several instruments from functional analysis to give the correct mathematical justification for the concept of "smoothness". What we can deduce in any case from the pseudo code are the three main elements, on which the RBF method is indeed based: An interpolating model (see Section 2.3.1), a (semi-)norm describing the smoothness (see Section 2.3.2) and the formulation of auxiliary problems (see Section 2.3.3). While the seminorm measures how "bumpy" an interpolating model is, the solutions of the auxiliary problems propose the next point to be evaluated, taking into account the "bumpiness" of the next interpolating model involving the next sample point. The choice of the next evaluation point is therefore a result of an optimization process, namely the minimization of the seminorm measuring the "bumpiness" of interpolations.

The following explanations mainly follow Gutmann's thesis [14].

### 2.3.1 The interpolating model

Assume that there are points $x^1, \ldots, x^n$ that are already evaluated and which are scattered in $R \subset \mathbb{R}^d$. The associated known function values are $f_i := f(x_i)$ where $f : \mathbb{R}^d \to \mathbb{R}$ is the expensive objective function. As it is common for surface methods, a model of the original objective function to the actually given data points

**Algorithm 2.1:** RBF method (pseudo code):

**Input**: Sample points $x^1, ..., x^n \in R$,

expensive function $f : \mathbb{R}^d \to \mathbb{R}$,

maximal number of evaluations $N$

**Output**: Sample point $\bar{x} \in R$ with best objective function value $\bar{f}$

**Initialization**: Compute $f_i := f(x^i)$ for $i = 1, \ldots, n$;

Set $\bar{x} \leftarrow \arg\min\{f_i\}$; $\bar{f} \leftarrow f(\bar{x})$; $k \leftarrow n$;

**while** $k \leq N$ **do**

    1. Compute the smoothest interpolating function $s(x)$ with

$$s(x^i) = f_i \quad \text{for } i = 1, \ldots, k.$$

    2. Compute the next point $y$ to be evaluated:

        a) Choose $f_k^*$ such that $f_k^* \leq s(x)$ for all $x \in R$.

        b) Choose $y \in R$ such that the new interpolating function $s_y(x)$ with

$$s_y(x^i) = f_i \quad \text{for } i = 1, \ldots, k$$
$$s_y(y) = f_k^*$$

        is *as smooth as possible*.

    3. Set $x^{k+1} := y$ and $f_{k+1} := f(x^{k+1})$ and update $(\bar{x}, \bar{f})$:

        $\bar{x} \leftarrow \arg\min\{f_1, \ldots, f_{k+1}\}$ and $\bar{f} \leftarrow f(\bar{x})$

    4. Set $k \leftarrow k + 1$.

**end**

has to be constructed at each stage of the algorithm. So the goal is to find a suitable interpolation, i.e. to find a function $s : \mathbb{R}^d \to \mathbb{R}$ with $s(x^i) = f(x^i)$ for all $i = 1, \ldots, n$. Of course, there is an infinite number of possibilities to achieve this aim.

In the onedimensional case the approach with the well-known natural cubic splines is obvious. Assume that $x^1 < \ldots < x^n$ are points in $\mathbb{R}$, then a natural cubic spline $s : \mathbb{R} \to \mathbb{R}$ is a function which can be represented (see e.g. [43]) as

$$s(x) = \sum_{i=1}^{n} \lambda_i |x - x^i|^3 + a + bx, \tag{2.1}$$

where the $n$ parameters $\lambda_1, \ldots, \lambda_n$ as well as the two parameters $a$ and $b$ are determined by a linear system of equations. This system of equations consists of the $n$ interpolation conditions

$$s(x^i) = f(x^i) \quad \text{for } i = 1, \ldots, n \tag{2.2}$$

and the two additional conditions

$$\sum_{i=1}^{n} \lambda_i = \sum_{i=1}^{n} \lambda_i x^i = 0. \tag{2.3}$$

The conditions (2.3) ensure that the second derivative of $s(x)$ vanishes at the endpoints $x^1$ and $x^n$, as it is common with natural cubic splines.

Conditions (2.3) will no longer be adequate when considering multidimensional functions. So we will reformulate conditions (2.3) as abstract conditions between the possible set of parameters $\lambda = (\lambda_1, \ldots, \lambda_n)$ and the space $\Pi_1^1$ of polynomials in one variable of degree 1. Since the monomials $\{1, x\}$ are a basis of $\Pi_1^1$, the inherent requirement of conditions (2.3) is that they hold for all elements of $\Pi_1^1$, i.e. conditions (2.3) are equivalent to the condition

$$\sum_{i=1}^{n} \lambda_i p(x^i) = 0 \quad \text{for all } p \in \Pi_1^1. \tag{2.4}$$

With both the interpolation conditions (2.2) and condition (2.4) the function $s(x)$ is uniquely determined. The well-known remarkable fact of natural splines is that they interpolate the given data, and moreover, they are the "smoothest" functions of all twice continuously differentiable functions that fulfill the interpolation conditions. The expression

$$\int_{\mathbb{R}} [f''(x)]^2 \, dx$$

11

is a common measure for the smoothness of a twice continuously differentiable function $f(x)$ of one variable. It is minimal for the natural cubic spline $s(x)$ compared to all twice continuously differentiable functions that fulfill the interpolation conditions above.

The interpolation of multidimensional data points can be derived from the onedimensional approach of natural cubic splines by generalizing two aspects. First, we replace the term $|x - x^i|^3$ in (2.1) by the general form $\phi(\|x - x^i\|)$, where $\phi(r) := r^3$ and $\|.\|$ denotes the Euclidean norm. Of course, we can choose a much more general function $\phi$; we will discuss the possible choices in Section 2.3.1.1. Second, we assume the polynomial $a + bx$ to be a general polynomial, i.e. we do no longer consider the space $\Pi_1^1$ of linear polynomials with the basis of monomials $p_1(x) := 1$ and $p_2(x) := x$. Instead we consider the space $\Pi_m^d$ of polynomials in $d$ variables of degree $m$ with a given basis of $\binom{d + m}{d} =: \hat{m}$ monomials $\{p_1(x), \ldots, p_{\hat{m}}(x)\}$. Then the adequate formulation of (2.1) for the multidimensional case is

$$s(x) = \sum_{i=1}^{n} \lambda_i \phi(\|x - x^i\|) + \sum_{j=1}^{\hat{m}} c_j p_j(x). \qquad (2.5)$$

Now the transfer of conditions (2.3) and (2.4) for the multidimensional case is technically straightforward. An equivalent formula for multidimensional functions of the two conditions (2.3) are the $\hat{m}$ conditions

$$\sum_{i=1}^{n} \lambda_i p_j(x^i) = 0 \quad \text{for } j = 1, \ldots, \hat{m} \qquad (2.6)$$

and the equivalent formula of condition (2.4) is

$$\sum_{i=1}^{n} \lambda_i p(x^i) = 0 \quad \text{for all } p \in \Pi_m^d. \qquad (2.7)$$

In Section 2.3.1.1 we discuss which functions $\phi : \mathbb{R}_+ \to \mathbb{R}$ are suitable for the approach in (2.5).

### 2.3.1.1 Radial basis functions and positive definiteness of functions

For the choice of $\phi$ in (2.5) we need functions that have suitable properties for the use of Gutmann's RBF-method. The commonly used functions are listed in Definition 2.1.

**Definition 2.1 (Radial basis functions)** *Let $\phi : \mathbb{R}_+ \to \mathbb{R}$ be one of the following functions, where $w \in \mathbb{N}_0$:*

1) surface splines: $\phi(r) = \begin{cases} r^\kappa & \text{with odd } \kappa \in \mathbb{N}, \\ r^\kappa \log r & \text{with even } \kappa \in \mathbb{N}, \end{cases}$

    *where special cases are*

      1 a) linear spline: $\kappa = 1$,

      1 b) thin plate spline: $\kappa = 2$,

      1 c) cubic spline: $\kappa = 3$,

2) multiquadrics: $\phi(r) = (r^2 + w^2)^\kappa$ *with* $\kappa > 0$ *and* $\kappa \notin \mathbb{N}$,

3) inverse multiquadrics: $\phi(r) = (r^2 + w^2)^\kappa$ *with* $\kappa < 0$,

4) Gaussians: $\phi(r) = \exp(-wr^2)$.

*We call each $\phi$ a* radial basis function.

For the further considerations we need the essential Definition 2.2 of functions being "conditionally positive definite".

**Definition 2.2 (Conditionally definite functions)** *Let $\sigma : \mathbb{R}^d \times \mathbb{R}^d \to \mathbb{R}$ be a function with $\sigma(x,y) = \sigma(y,x)$ for all $(x,y) \in \mathbb{R}^d \times \mathbb{R}^d$, i.e. let $\sigma$ be a symmetric function. Then $\sigma$ is called* conditionally positive definite of order $m+1$ *if*

$$\sum_{i,j=1}^{n} \lambda_i \lambda_j \sigma(x^i, x^j) > 0$$

*for any choice of pairwise different points $x^1, \ldots, x^n \in \mathbb{R}^d$ and any vector $\lambda \in \mathbb{R}^n \setminus \{0\}$ that satisfies*

$$\sum_{i=1}^{n} \lambda_i q(x^i) = 0 \qquad \text{for all } q \in \Pi_m^d. \tag{2.8}$$

*The function $\sigma$ is called* conditionally definite, *if either $\sigma$ or $-\sigma$ is conditionally positive definite.*
*A symmetric function $\sigma$ that is conditionally positive definite of order $0$ is called* positive definite.
*A function $\phi : \mathbb{R}_+ \to \mathbb{R}$ is called* conditionally positive definite of order $m+1$, *if the function $\sigma : \mathbb{R}^d \times \mathbb{R}^d \to \mathbb{R}$ defined by $\sigma(x,y) := \phi(\|x - y\|)$ is conditionally positive definite of order $m+1$ for every positive integer $d$. The function $\phi$ is called* conditionally definite, *if either $\phi$ or $-\phi$ is conditionally positive definite.*

Note that "conditional positive definiteness" for functions $\phi : \mathbb{R}_+ \to \mathbb{R}$ is independent of the number $d$.

In addition, the degree of conditional positive definiteness is a minimum degree, which is said in Remark 2.3.

**Remark 2.3** *Since $\Pi_k^d \subset \Pi_{k+l}^d$ for all $k, l \in \mathbb{N}$ we have that functions which are conditionally positive definite of order $k$ are also conditionally positive definite of order $k+l$ for all $l \in \mathbb{N}$.*

As a matter of fact, the radial basis functions from Definition 2.1 are conditionally positive definite of an appropriate order.

**Theorem 2.4 [14, Corollary 3.3]** *Let $\phi$ be one of the functions from Definition 2.1. If $m_\phi \in \mathbb{N}_0$ has the value*

$$m_\phi = \lfloor \tfrac{\kappa}{2} \rfloor + 1 \quad \text{(for surface splines)},$$
$$m_\phi = \lfloor \kappa \rfloor + 1 \quad \text{(for multiquadrics)},$$
$$m_\phi = 0 \quad \text{(for inverse multiquadrics or Gaussians)},$$

*then $(-1)^{m_\phi} \phi$ is conditionally positive definite of order $m_\phi$.*

The property of being conditionally positive definite will be important for computing an interpolating function (see Section 2.3.1.2) as well as for defining an appropriate measure of bumpiness (see Section 2.3.2).

### 2.3.1.2 A system of linear equations to be solved

We return to the interpolating model for the multidimensional case. With (2.6) we are able to formulate the system of linear equations that has to be solved for computing an interpolant (2.5). For this let $x^1, \ldots, x^n \in R \subset \mathbb{R}^d$ be the evaluated points so far and let $f_1, \ldots, f_n$ be the function values combined in a vector $F := (f_1, \ldots, f_n)^T$. Let $\phi$ be a radial basis function. Furthermore let $\{p_1(x), \ldots, p_{\hat{m}}(x)\}$ be an appropriate basis of monomials for the space $\Pi_m^d$ of polynomials. We define matrices $\Phi$ and $P$ as

$$\Phi_{ij} := \phi(\|x^i - x^j\|) \tag{2.9}$$

$$P_{ik} := p_k(x^i), \tag{2.10}$$

where $i = 1, \ldots, n$ and $j = 1, \ldots, n$ and $k = 1, \ldots, \hat{m}$.

With the definition of the vectors $\lambda$ resp. $c$ as

$$\lambda := (\lambda_1, \ldots, \lambda_n)^T \text{ resp. } c := (c_1, \ldots, c_{\hat{m}})^T \tag{2.11}$$

we finally get the system of linear equations as

$$\begin{pmatrix} \Phi & P \\ P^T & 0 \end{pmatrix} \begin{pmatrix} \lambda \\ c \end{pmatrix} = \begin{pmatrix} F \\ 0 \end{pmatrix} \begin{array}{l} \} \text{ I} \\ \} \text{ II} \end{array} \tag{2.12}$$

The system I represents the interpolation conditions (2.2) and system II represents the conditions (2.6). If we want (2.12) to provide an interpolant as stated in (2.5) we need to ensure that (2.12) has a solution.

Indeed there exists an assertion concerning the solution set of (2.12). To cite this, we first need Definition 2.5.

**Definition 2.5 (Unisolvency)** *A set $\{x^1, \ldots, x^n\} \subset \mathbb{R}^d$ is called $\Pi_m^d$ - unisolvent if*

$$q \in \Pi_m^d \text{ and } q(x^i) = 0 \text{ for all } i = 1, \ldots, n \text{ imply } q \equiv 0.$$

In the following we call $\Pi_m^d$ - unisolvent sets simply *unisolvent*. Unisolvency of a set of points $\{x^1, \ldots, x^n\}$ has an equivalent formulation in terms of the rank of $P$, with implications for the cardinality of the unisolvent set.

**Theorem 2.6 [14, Remark 3.5]** *A set $\{x^1, \ldots, x^n\}$ is unisolvent if and only if the system of equations $Pc = 0$ has only the solution $c = 0$, where $P$ and $c$ are defined in (2.10) and (2.11). This is equivalent to rank $P = \hat{m}$. Therefore a unisolvent set has at least $\hat{m}$ elements.*

**Proof:**
Let $\{p_1(x), \ldots, p_{\hat{m}}(x)\}$ be the basis of $\Pi_m^d$ that was used for building $P$ and let $q \in \Pi_m^d$ be represented as

$$q(x) = \sum_{j=1}^{\hat{m}} c_j p_j(x).$$

15

Then unisolvency is equivalent to

$$\sum_{j=1}^{\hat{m}} c_j p_j(x^1) = \ldots = \sum_{j=1}^{\hat{m}} c_j p_j(x^n) = 0 \quad \Rightarrow \quad c_1 = \ldots = c_{\hat{m}} = 0,$$

which is the same as to say

$$Pc = 0 \quad \Rightarrow \quad c = 0.$$

$\square$

The essential assertion about the system of equations (2.12) is that it has a unique solution for an appropriate choice of interpolation points.

**Theorem 2.7 [14, Theorem 3.6]** *Let $\phi$ be a function from Definition 2.1, let $x^1, \ldots, x^n \in \mathbb{R}^d$ be pairwise different points, and let $\Phi$ and $P$ be defined as in (2.9) and (2.10), respectively, where we assume that $m \geq m_\phi - 1$ for the considered set $\Pi_m^d$. Then the interpolation matrix*

$$A := \begin{pmatrix} \Phi & P \\ P^T & 0 \end{pmatrix} \in \mathbb{R}^{(n+\hat{m}) \times (n+\hat{m})}$$

*is nonsingular if and only if $\{x^1, \ldots, x^n\}$ is unisolvent.*

In the case of natural cubic splines we have seen that not only there exists a unique solution of the system of linear equations, but that the resulting interpolant $s$ also has the property of being the "smoothest" function within a special set of functions. In fact, there exists a similar theorem for the other radial basis functions concerning an abstract function space, called $\mathcal{A}_{\phi,m}(\Omega)$, which we will introduce in Section 2.3.2.

### 2.3.2 A measure of bumpiness

Now the goal is to find a measure that, similar to the natural cubic spline case, describes the "bumpiness" of the multidimensional interpolant (2.5). Since it is very difficult to imagine any formal measure for multidimensional functions, it will be helpful to take a look at the remarkable relation in the special case of the onedimensional natural cubic splines [14, 43]. For this let us assume, that the assumptions of Theorem 2.7 are fulfilled, where $\phi(r) = r^3$ and $d = 1$, $\Phi$ is given by equations (2.9) and $\lambda$ is the solution of system (2.12). Then we have the following relation between an expression of a onedimensional function on the one hand and an expression, which consists mainly of a matrixproduct, on the other hand:

$$\int_{\mathbb{R}} [s''(x)]^2 \, dx = 12\lambda^T \Phi \lambda. \tag{2.13}$$

To see (2.13) let $x^1 < \ldots < x^n \in \mathbb{R}$ and let $s(x) = \sum_{i=1}^n \lambda_i |x - x_i|^3 + a + bx$ be the unique interpolant determined by the conditions (2.2) and (2.3). Then we can write $s(x)$ and the first three derivatives of $s(x)$ for $x \in [x^t, x^{t+1}]$ as

$$s(x) = \sum_{i=1}^t \lambda_i (x - x_i)^3 - \sum_{i=t+1}^n \lambda_i (x - x_i)^3 + a + bx,$$

$$s'(x) = 3 \sum_{i=1}^t \lambda_i (x - x_i)^2 - 3 \sum_{i=t+1}^n \lambda_i (x - x_i)^2 + b,$$

$$s''(x) = 6 \sum_{i=1}^t \lambda_i (x - x_i) - 6 \sum_{i=t+1}^n \lambda_i (x - x_i),$$

$$s'''(x) = 6 \sum_{i=1}^t \lambda_i - 6 \sum_{i=t+1}^n \lambda_i = 6(\sum_{i=1}^t \lambda_i - \sum_{i=t+1}^n \lambda_i) = 6(2 \sum_{i=1}^t \lambda_i) = 12 \sum_{i=1}^t \lambda_i,$$

where in the last row we used the condition $\sum_{i=1}^n \lambda_i = 0$ from (2.3).

Using integration by parts we get

$$
\begin{aligned}
\int_{\mathbb{R}} [s''(x)]^2 dx &= s''(x)s'(x)\Big]_{x_1}^{x_n} - \int_{x_1}^{x_n} s'''(x)s'(x)dx \\
&= \quad 0 \quad - \sum_{i=1}^{n-1} \int_{x_i}^{x_{i+1}} s'''(x)s'(x)dx \\
&= -\sum_{i=1}^{n-1} \int_{x_i}^{x_{i+1}} 12 \sum_{k=1}^{i} \lambda_k s'(x)dx \\
&= -12 \sum_{i=1}^{n-1} \sum_{k=1}^{i} \lambda_k \int_{x_i}^{x_{i+1}} s'(x)dx \\
&= -12 \sum_{i=1}^{n-1} \sum_{k=1}^{i} \lambda_k \left[ s(x^{i+1}) - s(x^i) \right] \\
&= 12 \sum_{i=1}^{n-1} \sum_{k=1}^{i} \lambda_k \left[ s(x^i) - s(x^{i+1}) \right] \text{ (telescopic sum)} \\
&= 12 \left[ \sum_{i=1}^{n-1} \lambda_i s(x^i) - \sum_{i=1}^{n-1} \lambda_i s(x^n) \right] \\
&= 12 \left[ \sum_{i=1}^{n-1} \lambda_i s(x^i) + \lambda_n s(x^n) \right] \\
&= 12 \sum_{i=1}^{n} \lambda_i s(x^i).
\end{aligned}
$$

where we again used $\sum_{i=1}^{n} \lambda_i = 0$ in the last but one equation.

Using again conditions (2.3) and $\Phi_{ij} = |x^i - x^j|^3$ we get

$$
\begin{aligned}
12 \sum_{i=1}^{n} \lambda_i s(x^i) &= 12 \sum_{i=1}^{n} \lambda_i (\sum_{j=1}^{n} \lambda_j |x^i - x^j|^3 + a + bx^i) \\
&= 12(\sum_{i,j=1}^{n} \lambda_i \lambda_j \Phi_{ij} + a \sum_{i=1}^{n} \lambda_i + b \sum_{i=1}^{n} \lambda_i x^i) \\
&= 12 \sum_{i,j=1}^{n} \lambda_i \lambda_j \Phi_{ij} \\
&= 12 \lambda^T \Phi \lambda,
\end{aligned}
$$

which finally shows (2.13).

We repeat the following fact: The formula on the left hand side of (2.13) measures the "smoothness" of the optimal natural cubic spline $s$ and its value for the natural cubic spline $s$ is minimal within the class of twice continuously differentiable functions that fulfill the interpolation conditions. In addition, we see that within the formula on the right hand side of (2.13) the expression $\lambda^T \Phi \lambda$ is positive due to the fact, that $\int_{\mathbb{R}} [s''(x)]^2 \, dx$ is positive except for $s \equiv 0$. Therefore $\lambda^T \Phi \lambda$ can be seen as a measure of bumpiness likewise. The advantage of the last expression is its independence of dimension. Therefore it would be beneficial if the expression is positive for the other radial basis functions as well. Indeed, as a direct result of Theorem 2.4 and the definition of $\Phi$ we have that

$$(-1)^{m_\phi} \lambda^T \Phi \lambda > 0, \tag{2.14}$$

if $\lambda \in \mathbb{R}^n$ is any nonzero vector that satisfies (2.8) for $m = m_\phi - 1$. Therefore the nonnegativity of $(-1)^{m_\phi} \lambda^T \Phi \lambda$ is guaranteed and can be used for further considerations.

As indicated in the last section, the generalization of natural cubic splines for the multidimensional case is very abstract. Therefore a corresponding "smoothness"-property is not easily available, nor is a class of functions for which this property of the interpolant applies. Thus, it is useful to take a look at the space of functions with the same components as the interpolant $s$: The first part is a sum of weighted radial basis functions, where the weights $\lambda_1, \ldots, \lambda_n$ fulfill condition (2.7) with respect to the polynomial space $\Pi_m^d$. For a given general set $\Omega \subseteq \mathbb{R}^d$ we define the function space

$$\mathcal{F}_{\phi,m}(\Omega) := \{ f : \mathbb{R}^d \to \mathbb{R} : f(x) = \sum_{i=1}^n \lambda_i \phi(\|x - x^i\|) \text{ with } \lambda_1, \ldots, \lambda_n \in \mathbb{R},$$

$$x^1, \ldots, x^n \in \Omega \text{ such that } \sum_{i=1}^n \lambda_i q(x^i) = 0 \text{ for all } q \in \Pi_m^d \}$$

The second part of the functions is given by a polynomial from $\Pi_m^d$ and we finally define the function space

$$\mathcal{A}_{\phi,m}(\Omega) := \mathcal{F}_{\phi,m}(\Omega) + \Pi_m^d.$$

It is obvious from (2.5) that $s \in \mathcal{A}_{\phi,m}(\Omega)$.

With the two elements of $\mathcal{A}_{\phi,m}(\Omega)$,

$$s(x) = \sum_{i=1}^{N(s)} \lambda_i \phi(\|x - y^i\|) + p(x)$$

and

$$u(x) = \sum_{i=1}^{N(u)} \mu_i \phi(\|x - z^i\|) + q(x),$$

we can define (see [14]) a semi-inner product

$$\langle s, u \rangle := (-1)^{m_\phi} \sum_{i=1}^{N(s)} \lambda_i u(y^i)$$

on $\mathcal{A}_{\phi,m}(\Omega)$. The positive integer $m_\phi$ is defined in Theorem 2.4 and depends on the chosen radial basis function $\phi$.

We prove that $\langle \cdot, \cdot \rangle$ is indeed a semi-inner product by verifying symmetry, bilinearity and nonnegativity:

1. Symmetry: We have

$$\langle s, u \rangle = (-1)^{m_\phi} \sum_{i=1}^{N(s)} \lambda_i \left( \sum_{j=1}^{N(u)} \mu_j \phi(\|y^i - z^j\|) + q(y^i) \right)$$

$$= (-1)^{m_\phi} \sum_{i=1}^{N(s)} \sum_{j=1}^{N(u)} \lambda_i \mu_j \phi(\|y^i - z^j\|)$$

$$= (-1)^{m_\phi} \sum_{j=1}^{N(u)} \mu_j \left( \sum_{i=1}^{N(s)} \lambda_i \phi(\|z^j - y^i\|) + p(z^j) \right)$$

$$= (-1)^{m_\phi} \sum_{j=1}^{N(u)} \mu_j s(z^j)$$

$$= \langle u, s \rangle,$$

where we used $\sum_{i=1}^{N(s)} \lambda_i q(y^i) = 0$ and $\sum_{j=1}^{N(u)} \mu_j p(z^j) = 0$ in rows two and three, respectively.

2. Bilinearity:
   Obviously $\langle s, u \rangle$ is linear in its second argument and symmetric. Therefore, $\langle s, u \rangle$ is also linear in its first argument and therefore bilinear.

3. Nonnegativity: We have

$$\langle s, s \rangle = (-1)^{m_\phi} \sum_{i=1}^{N(s)} \lambda_i s(y^i)$$

$$= (-1)^{m_\phi} \sum_{i,j=1}^{N(s)} \lambda_i \lambda_j \phi(\|y^i - y^j\|) + \sum_{i=1}^{N(s)} \lambda_i p(y^i)$$

$$= (-1)^{m_\phi} \sum_{i,j=1}^{N(s)} \lambda_i \lambda_j \phi(\|y^i - y^j\|)$$

$$\geq 0,$$

where we used again that $\sum_{i=1}^{N(s)} \lambda_i p(y^i) = 0$. Nonnegativity in the last row follows from the conditional positive definiteness of the function $(-1)^{m_\phi}\phi$ (see Theorem 2.4).

If we assume that $\langle s, s \rangle = 0$, then the conditional definiteness of $\phi$ implies $\lambda_i = 0$ for $i = 1, \ldots, N(s)$ and therefore we have $s \in \Pi_m^d$. Thus the null space of $\langle \cdot, \cdot \rangle$ is $\Pi_m^d$ (see [14]).

The definition of $\langle \cdot, \cdot \rangle$ induces the seminorm

$$\| \cdot \| := \langle \cdot, \cdot \rangle^{\frac{1}{2}}.$$

According to this definition the seminorm $\|s\|$ for an element $s \in \mathcal{A}_{\phi,m}(\Omega)$ with the $N(s)$ interpolation points $x^1, \ldots, x^{N(s)}$ is

$$\|s\| = \sqrt{(-1)^{m_\phi} \sum_{i=1}^{N(s)} \lambda_i s(x^i)}$$

$$= \sqrt{(-1)^{m_\phi} \sum_{i,j=1}^{N(s)} \lambda_i \lambda_j \phi(\|x^i - x^j\|)}$$

$$= \sqrt{(-1)^{m_\phi} \lambda^T \Phi \lambda},$$

where the polynomial part of $s(x^i)$ vanishes for the same reason as in the proof of nonnegativity of the semi-inner product. The positivity of the last expression is guaranteed by (2.14), derived from Theorem 2.4.

With this seminorm we are able to state the central assertion concerning the extremal property of the multidimensional interpolant determined by the system of linear equations (2.12).

**Theorem 2.8 [14, Theorem 3.7]** *Let $\phi$ be any radial basis function and let $m$ be chosen such that $m \geq m_\phi - 1$. Furthermore, let $\Omega \subset \mathbb{R}^d$, and assume that a unisolvent set $\{x^1, \ldots, x^n\} \subset \Omega$ and values $f_1, \ldots, f_n \in \mathbb{R}$ are given. Let $s$ be the interpolant whose coefficients solve (2.12). Then $s$ is the unique element of $\mathcal{A}_{\phi,m}(\Omega)$ that minimizes the seminorm $\|g\|$ on the set*

$$\left\{ g \in \mathcal{A}_{\phi,m}(\Omega) : g(x^i) = f_i \text{ for all } i = 1, \ldots, n \right\}.$$

Theorem 2.8 describes the extremal behaviour of $s$ on the space $\mathcal{A}_{\phi,m}(\Omega)$. In the case of onedimensional natural cubic splines the extreme property, i.e. the maximal "smoothness", was proven over the large class of twice differentiable functions, which is much less abstract compared to the space $\mathcal{A}_{\phi,m}(\Omega)$. It is desirable to extend the set $\mathcal{A}_{\phi,m}(\Omega)$ in Theorem 2.8 to a larger class of analytically describable functions. Indeed there exists a larger class of functions called the "native space", in which the space $\mathcal{A}_{\phi,m}(\Omega)$ is embedded and for which Theorem 2.8 holds as well (see [14]). We will give an example of the native space in case of "natural cubic splines" at the end of section 2.3.3. However, for our purposes there is no need for more detailed function spaces.

We finish this section with a result that shows the importance of the introduced seminorm for the sequence of interpolants within the RBF method. As a consequence of Theorem 2.8, the seminorm of the sequence of interpolants increases with a growing set of interpolation points:

**Corollary 2.9 [14, Corollary 3.8]** *Let $\phi, m$ and $\Omega$ be as in Theorem 2.8, and let $n, k \in \mathbb{N}$. Assume that pairwise different points $x^1, \ldots, x^{n+k} \in \Omega$ and values $f_1, \ldots, f_{n+k} \in \mathbb{R}$ are given, and that $\{x^1, \ldots, x^n\}$ is unisolvent. Furthermore, let $s_n$ be the optimal interpolant to $(x^1, f_1), \ldots, (x^n, f_n)$ and let $s_{n+k}$ be the optimal interpolant to $(x^1, f_1), \ldots, (x^{n+k}, f_{n+k})$. Then*

$$\| s_{n+k} - s_n \|^2 = \| s_{n+k} \|^2 - \| s_n \|^2.$$

*It follows that $\| s_n \| \leq \| s_{n+k} \|$.*

## 2.3.3 Auxiliary problems for the next sample point

After the construction of an interpolating model described in Section 2.3.1 as well as deducing a convenient seminorm in Section 2.3.2 the next and main step in the RBF method is the choice of the sample point, step 2 in the pseudo code of Algorithm 2.1, where the function $f$ ought to be evaluated next. Here the introduced seminorm plays the essential part. To distinguish the interpolants at different stages of the algorithm, we will henceforth denote the interpolants with $s_n$ instead of $s$, where the number $n$ indicates the number of interpolation points. Now the main idea within the method is the hypothesis that a value $f_n^*$ is the value of the objective function $f$ at an (unknown) point $y$. Now let us assume that in Step 2 of the pseudocode we have $n$ interpolation points $x^1, \ldots, x^n$ with the appropriate function evaluations

$f_i$ and a value $f_n^*$. Then the task is to find the interpolant $s_y(x)$ that fulfills the restrictions

$$s_y(x^i) = f_i \quad \text{for all } i = 1, \ldots, n$$
$$s_y(y) = f_n^*. \tag{2.15}$$

Note that $y$ is not known so far. So in the further considerations $y$ will be treated like a variable in a special optimization problem. The optimization problem is to find that $y$ for which the associated function $s_y$ minimizes the seminorm introduced in Section 2.3.2.

Let $l_n(y, x) \in \mathcal{A}_{\phi,m}(\Omega)$ be the optimal interpolant which fulfills

$$l_n(y, x^i) = 0 \quad \text{for all } i = 1, \ldots, n \tag{2.16}$$
$$l_n(y, y) = 1. \tag{2.17}$$

Due to the fact that both $l_n(y, x)$ and $s_n(x)$ are elements of the set $\mathcal{A}_{\phi,m}(\Omega)$ we can express $s_y$ that fulfills (2.15) as:

$$s_y(x) = s_n(x) + [f_n^* - s_n(y)]l_n(y, x).$$

Using these notations we can write the function $l_n(y, x)$ as the sum of radial basis functions and a polynomial:

$$l_n(y, x) = \sum_{i=1}^{n} \alpha_i(y)\phi(\|x - x^i\|) + \beta(y)\phi(\|x - y\|) + \sum_{i=1}^{\hat{m}} b_i(y)p_i(x),$$

where $\alpha(y) := (\alpha_1(y), \ldots, \alpha_n(y))$ and $\beta(y)$ as well as $b(y) := (b_1(y), \ldots, b_{\hat{m}}(y))$ depend on $y$.

In addition, corresponding to conditions (2.6), $l_n(y, x)$ has to fulfill the conditions

$$\sum_{i=1}^{n} \alpha_i(y)p_j(x^i) + \beta(y)p_j(y) = 0 \text{ for all } j = 1, \ldots, \hat{m}. \tag{2.18}$$

With the definition of the vectors

$$\pi(y) := (p_1(y), \ldots, p_{\hat{m}}(y))^T \tag{2.19}$$

and

$$u^n(y) := (\phi(\|x^1 - y\|), \ldots, \phi(\|x^n - y\|))^T \tag{2.20}$$

we can rewrite (2.16), (2.17) and (2.18) as the following system of linear equations

$$\underbrace{\begin{pmatrix} \Phi & u^n(y) & P \\ u^n(y)^T & \phi(0) & \pi(y)^T \\ P^T & \pi(y) & 0 \end{pmatrix}}_{=:A(y)} \begin{pmatrix} \alpha(y) \\ \beta(y) \\ b(y) \end{pmatrix} = \begin{pmatrix} 0_n \\ 1 \\ 0_{\hat{m}} \end{pmatrix}.$$

Only the value $\beta(y)$ is needed in the further considerations. With the help of Cramer's rule the calculation of $\beta(y)$ is reduced to

$$\beta(y) = \frac{\det A}{\det A(y)},$$

where $A$ is the interpolation matrix $\begin{pmatrix} \Phi & P \\ P^T & 0 \end{pmatrix}$.

Furthermore, due to (2.16), we have that

$$\langle\, s_n, l_n(y, \cdot)\,\rangle = (-1)^{m_\phi} \sum_{i=1}^{n} \lambda_i l_n(y, x_i) = 0,$$

which leads to the fact that the seminorm of $s_y$ is related to the seminorm of $l_n(y, \cdot)$ by the equation

$$\|s_y\|^2 = \|s_n\|^2 + [f_n^* - s_n(y)]^2 \|l_n(y, \cdot)\|^2.$$

Knowing that $\|s_n\|^2$ does not depend on $y$ we have as a result that minimizing the bumpiness $\|s_y\|^2$ of $s_y$ corresponds to solving

$$\min g_n(y) := [f_n^* - s_n(y)]^2 \|l_n(y, \cdot)\|^2 \text{ s.t. } y \in R\backslash\{x^1, \ldots, x^n\} \tag{2.21}$$

where the problem

$$\min \mu_n(y) := \|l_n(y, \cdot)\|^2 \text{ s.t. } y \in R\backslash\{x^1, \ldots, x^n\} \tag{2.22}$$

is of special interest.

The squared seminorm of $l_n(y, \cdot)$ is

$$\|l_n(y, \cdot)\|^2 = (-1)^{m_\phi} \left[ \sum_{i=1}^{n} \alpha_i(y) l_n(y, x^i) + \beta(y) l_n(y, y) \right] \tag{2.23}$$
$$= (-1)^{m_\phi} \beta(y),$$

where the last equation comes from conditions (2.16) and (2.17).

Because of the equation (2.23) the explicit formulation of the minimization problem (2.21) is

$$\min g_n(y) = (-1)^{m_\phi} [f_n^* - s_n(y)]^2 \frac{\det A}{\det A(y)} \text{ s.t. } y \in R\backslash\{x^1, \ldots, x^n\} \tag{2.24}$$

and the explicit formulation of the minimization problem (2.22) is

$$\min \mu_n(y) = (-1)^{m_\phi} \frac{\det A}{\det A(y)} \text{ s.t. } y \in R\backslash\{x^1, \ldots, x^n\}. \tag{2.25}$$

The two auxiliary problems (2.24) and (2.25) represent two different search strategies, which are both necessarily employed within global optimization. Problem (2.24) incorporates a local search strategy. Here the interpolant $s_n$ influences the optimization process more or less, depending on the choice of $f_n^*$. The extreme case $f_n^* = \min_{x \in R} s_n(x)$ is a good choice, if $s_n$ is a good model for the expensive objective. The smaller we choose $f_n^*$ the more the influence of $s_n$ disappeares in the expression $[f_n^* - s_n(y)]$ and therefore in the search strategy. The extreme case in the opposite is the smallest possible $f_n^*$, which we call $f_n^* = -\infty$. This corresponds to treating problem (2.25). The optimization of (2.25) does not depend on $s_n$ but on the matrices $A$ and $A(y)$ only and therefore mainly on the position of the sample points $x^1, \ldots, x^n$. The minimization of (2.25) represents a global search strategy because here the solution $x^{n+1}$ lies as far away as possible from the other sample points (see [14]). Due to the fact that the quality of the interpolant $s_n$ is not known at any stage of the algorithm, it is recommended to alternately solve (2.24) and (2.25) and, in addition, to vary the value $f_n^*$. Gutmann presents a few so-called "cycles" in his dissertation.

Although the RBF method generates good results in general, convergence results are scarce. The convergence of the RBF method was only shown for the *surface splines* in Definition 2.1 (see [14]). The convergence results indicate that it is advantageous if the objective lies in the so-called "Native space" of the chosen radial basis function $\phi$. In shortform the "Native space" is the set of functions $\{f : \Omega \to \mathbb{R}\}$ where for any unisolvent set $\{x^1, \ldots, x^n\}$ the seminorm of the optimal interpolant $s$ for each $f$ is smaller than a constant $C_f$, where the subscript $f$ means that the constant depends only on $f$. The formal definition of $\mathcal{N}_{\phi,m}(\Omega)$ is the following: For a given set $\Omega \subseteq \mathbb{R}^d$ the set of functions

$$\mathcal{N}_{\phi,m}(\Omega) := \{f : \Omega \to \mathbb{R} \mid \|s\| \le C_f \forall\, s \in \mathcal{A}_{\phi,m}(\Omega) \text{ with } \{x_1, \ldots, x_n\} \subset \Omega \text{ unisolvent}\}$$

is called *Native Space for $\phi$ and $m$*. In the onedimensional case of "natural cubic splines", i.e. $\phi(r) = r^3$ and $m = d = 1$, this Native space coincides with the class of twice continuously differentiable functions [40, 43], i.e. for $[a, b] \subseteq \mathbb{R}$ we have

$$\mathcal{N}_{\phi,m}([a, b]) = W_2^2[a, b]$$

with $W_2^2[a, b] = \{f : f'' \in L_2[a, b]\}$, i.e. $W_2^2[a, b] = \{f : \int_a^b |f''(x)|^2 dx < \infty\}$, and $\|.\|_{L_{2[a,b]}}$ is introduced by the semi inner product

$$\langle f'', g'' \rangle_{L_2[a,b]} = \int_a^b f''(x)g''(x)dx.$$

For the other radial basis functions there does not exist such a convenient analytical representation of the native space. Therefore the decision, whether a function $f$

belongs to one of these native spaces, is much more difficult than in the case of natural cubic splines. The exploration of "Native spaces" is one important research topic, as it is said in [41]: "It is one of the most challenging research topics to deduce properties of the native space from properties of $\sigma$."

As Gutmann proves in his article [13], the two auxiliary problems (2.24) and (2.25) are discontinuous at the points $x^1, \ldots, x^n$. In the following considerations we will reformulate them into continuous problems in the same way as Gutmann did in his article. It is a challenge to determine the global optimum of the two (reformulated) problems (2.24) and (2.25) because the objective functions are multimodal. Since the success of the whole algorithm depends on how accurately the solutions for these subproblems are computed, it is preferable to solve these problems exactly.

# Chapter 3

# An optimized choice of points to be evaluated

In recent years Gutmann's powerful RBF methods have inspired other scientists to develop algorithms including (modified) parts of his original approach. The modifications concern the choice of $f^*$ ([18]), efficient solutions of the systems of equations (2.12) ([6]), the weighing of local vs. global search for chosing the next point to be evaluated ([39]) or the treatment of expensive restrictions as penalty functions ([6, 18]), for example. Furthermore, some of the authors (for example [6]), and even Gutmann himself, mention the importance of exact solutions for the original subproblems (2.24) and (2.25), which are formulated in Section 2.3.3. In the following sections we will focus on improving the RBF method with regard to this aspect.

This chapter is divided into five parts. First of all, in Section 3.1, we will reformulate the original subproblems (2.25) respectively (2.24) into the continuous *auxiliary problem* respectively *weighted auxiliary problem*. In addition, we will specify the matrices and parameters that occur within the problems. This also includes the specific form of the interpolating function $s(x)$ due to its dependence on the specific matrices and parameters. After that, in Section 3.2, we will go on with specifying special properties of the radial basis functions $\phi$ and, in the first instance, computing lower bounds for the interpolating function $s(x)$. The necessity for this will be explained. In Section 3.3 we lay the foundations for computing lower bounds for the *auxiliary problem* respectively *weighted auxiliary problem*, i.e. we will propose theoretical results concerning the structure of the occurring interpolation matrices $\Phi$ and $A$ respectively $\Phi^{-1}$ and $A^{-1}$. Based on this, we will formulate the lower bounds for the *auxiliary problem* respectively *weighted auxiliary problem* in Section 3.4. Finally, in Section 3.5, we will propose the Branch and Bound algorithm to solve these subproblems.

## 3.1 The subproblems: Cheap global optimization problems

As stated at the end of Section 2.3.3 the objective functions of the two auxiliary problems (2.24) and (2.25) are discontinuous at the points $x^1, \ldots, x^n$. Before we reformulate these problems into continuous problems we will recall some notation. For given points $x^1, \ldots, x^n \in \mathbb{R}^d$, a given basis function $\phi$ and a given basis $p_1(x), \ldots, p_{\hat{m}}(x)$ of the space of polynomials $\Pi_m^d$ we define the vectors

$$u^n(x) := (\phi(\|x^1 - x\|), \ldots, \phi(\|x^n - x\|))^T \tag{3.1}$$

and

$$\pi(x) := (p_1(x), \ldots, p_{\hat{m}}(x))^T$$

like in (2.20) and (2.19). We combine them into the vector

$$z^n(x) := (u^n(x)^T, \pi(x)^T)^T.$$

The matrix

$$A := \begin{pmatrix} \Phi & P \\ P^T & 0 \end{pmatrix}$$

is constructed as in (2.9) and (2.10).

In Remark 3.1 we will give a complete overview of the matrices $A$, vectors $z_n(x)$ and polynomial degree $m$ which depend on the chosen radial basis function $\phi$. Observe that the degree $m$ determines the size $\hat{m}$ of the basis of $\Pi_m^d$ and therefore the size of the matrix $P$. The matrix $P$ in turn is contained in the matrix $A$. We will use the minimal possible degree of $m$, which is equal to $m_\phi - 1$ (see Theorem 2.7), for each indiviual basis function $\phi$.

**Remark 3.1** *The explicit values of $m_\phi$ (and $m$ and $\hat{m}$), $A$ and $z^n(x)$ in Definitions 3.3 and 3.6 are different respectively can be chosen to be different for every type of radial basis function (see also [13]). We will use the following values. (Here $x_j^i$ means the $j$-th entry of $x^i$ and $x_j$ means the $j$-th entry of $x$.)*

i) *inverse multiquadric, Gaussians: $m = m_\phi - 1 = -1$, $\hat{m} = 0$ and $A = \Phi$*
   *and $z^n(x) = u^n(x)$ ,*

**ii)** *multiquadrics with $\kappa < 1$, surface spline with $\kappa = 1$: $m = m_\phi - 1 = 0$, $\hat{m} = 1$ and*

$$A = \begin{pmatrix} \Phi & \mathbb{1} \\ \mathbb{1}^T & 0 \end{pmatrix}$$

*and $z^n(x) = (u^n(x)^T, 1)^T$. Here $\mathbb{1}$ denotes the vector of ones.*

**iii)** *surface splines with $\kappa = 2, 3$: $m = m_\phi - 1 = 1$, $\hat{m} = d + 1$, where we have $d + 1$ monomials because the dimension $\hat{m}$ of the space $\Pi_m^d$ with $m = 1$ is $\hat{m} = \binom{d+1}{d} = d + 1$, and*

$$A = \begin{pmatrix} \Phi & P \\ P^T & 0 \end{pmatrix},$$

$$P = \begin{pmatrix} p_1(x^1) & \ldots & p_{d+1}(x^1) \\ \ldots & \ldots & \ldots \\ p_1(x^n) & \ldots & p_{d+1}(x^n) \end{pmatrix} = \begin{pmatrix} 1 & x_1^1 & \ldots & x_d^1 \\ 1 & x_1^2 & \ldots & x_d^2 \\ \vdots & \vdots & \vdots & \vdots \\ 1 & x_1^n & \ldots & x_d^n \end{pmatrix}$$

*and $z^n(x) = (u^n(x)^T, \pi(x)^T)^T$, with $\pi(x) = (1, x_1, \ldots, x_d)$.*

The interpolating function $s_n$ depends, like the matrices and vectors in Remark 3.1, on the chosen function $\phi$ and the parameter $\hat{m}$. Therefore we are able to list the interpolating functions $s_n$ in a very similar way as in Remark 3.1. The parameters $(\lambda, c)^T$ of $s_n(x)$ are the solutions of (2.12). Depending on the chosen $\phi$ the function $s_n$ looks as follows:

**Remark 3.2** *The explicit interpolating function $s_n(x)$ for each individual basis function $\phi$ for given interpolation points $x^1, \ldots, x^n$ is as follows.*

**i)** inverse multiquadric, Gaussians ($\hat{m} = 0$, *i.e. no polynomial* ):

$$s_n(x) = \sum_{i=1}^n \lambda_i \phi(\|x - x^i\|)$$

**ii)** multiquadrics *with $\kappa < 1$, surface spline with $\kappa = 1$ ($\hat{m} = 1$, i.e. constant polynomial* ):

$$s_n(x) = \sum_{i=1}^n \lambda_i \phi(\|x - x^i\|) + c$$

**iii)** surface splines *with $\kappa = 2, 3$ ($\hat{m} = d + 1$, i.e. linear polynomial* ):

$$s_n(x) = \sum_{i=1}^n \lambda_i \phi(\|x - x^i\|) + c_1 + \sum_{j=2}^{d+1} c_j x_{j-1}.$$

Using Remarks 3.1 and 3.2 we will reformulate the problems (2.24) and (2.25) as continuous problems (see [14]).

For (2.25) let us consider the following Definition 3.3.

**Definition 3.3** *For a given box $[x^L, x^U] \subset \mathbb{R}^d$, the problem*

$$\min a_n(x) := (-1)^{m_\phi} \left[ z^n(x)^T A^{-1} z^n(x) - \phi(0) \right] \quad s.t. \ x \in \left[ x^L, x^U \right] \tag{3.2}$$

*is called* auxiliary problem.

The important relation between the function $a_n$ and the function $\mu_n$ we defined in Section 2.3.3 is given in Lemma 3.4

**Lemma 3.4** *We have*

**1)** $a_n(y) = -\frac{1}{\mu_n(y)}$ *for all* $y \in R \backslash \{x^1, \ldots, x^n\}$ *and*

**2)** $a_n(y) \leq 0$ *for all* $y \in R \backslash \{x^1, \ldots, x^n\}$.

**Proof:** Part 1) follows from [14, Proposition 4.12]. Part 2) follows from 1) and (2.22).

□

With Lemma 3.5 we finally give Definition 3.3 its justification.

**Lemma 3.5** *The continuous problem (3.2) has the same minimization points as the discontinuous problem (2.25) restricted to the box* $\left[ x^L, x^U \right] \backslash \{x^1, \ldots, x^n\}$.

**Proof:** Follows from Lemma 3.4.

□

For (2.24) let us consider Definition 3.6.

**Definition 3.6** *Let $f^*$ be a real value with $f^* \leq s_n(x) \, \forall x \in R$. Then the problem*

$$\min w_n(x) := (-1)^{m_\phi} \frac{1}{[s_n(x) - f^*]^2} \left[ z^n(x)^T A^{-1} z^n(x) - \phi(0) \right] \quad s.t. \ x \in \left[ x^L, x^U \right] \tag{3.3}$$

*is called* weighted auxiliary problem.

Here a relation between the function $w_n$ and the function $g_n$ exists as well and is given in Lemma 3.7.

**Lemma 3.7** *We have*

**1)** $w_n(y) = -\frac{1}{g_n(y)}$ *for all* $y \in R \backslash \{x^1, \dots, x^n\}$ *and*

**2)** $w_n(y) \leq 0.$

**Proof:** Part 1) follows from [14, Proposition 4.12]. Part 2) follows from 1) and (2.21).

$\square$

With Lemma 3.8 Definition 3.6 becomes reasonable.

**Lemma 3.8** *The continuous problem (3.3) has the same minimization points as the discontinuous problem (2.24) restricted to the box* $\left[x^L, x^U\right] \backslash \{x_1, \dots, x_n\}.$

**Proof:** Follows from Lemma 3.7.

$\square$

## 3.2 Distances, properties of $\phi(r)$ and lower bounds for $s_n(x)$

As we have seen in the pseudocode of the RBF method in Section 2.3 we need the number $f_n^*$ as a lower bound of the function $s_n(x)$ on $R$. To get such an $f_n^*$ we have to solve the global optimization problem

$$\min s_n(x), \quad s.t. \quad x \in R.$$

We will solve this problem by using a branch and bound algorithm, which will be proposed in Section 3.5. Therefore we need to compute lower bounds of $s_n(x)$ on subrectangles of $R$. We will use some nice properties of the radial basis function $\phi$ to compute these bounds. As it is common for branch and bound routines, we will consider sequences of boxes. It will be useful to define special distances of these boxes to every single sample point. As we will solve the *auxiliary problem* and the *weighted auxiliary problem* also with a branch and bound routine, the Section 3.2.1 will be needed for Section 3.4 as well.

### 3.2.1 Minimal and maximal distances

**Definition 3.9** *Let $x^i \in \{x^1, \ldots, x^n\}$ and let $R \subset \mathbb{R}^d$ be a rectangle. Then*

$$x_{min}^{iR} := \operatorname{argmin} \left\{ \|x^i - x\| : x \in R \right\}$$

*resp.*

$$x_{max}^{iR} := \operatorname{argmax} \left\{ \|x^i - x\| : x \in R \right\}$$

*defines those points in $R$ which are the nearest resp. farthest from $x^i$.*

*With $d^{iR} := \|x^i - x_{min}^{iR}\|$ resp. $D^{iR} := \|x^i - x_{max}^{iR}\|$ we denote the* minimum *resp.* maximum *distance of $x^i$ to $R$.*

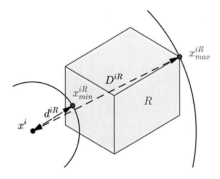

Figure 3.1: Minimum resp. maximum distance between $x^i$ and the rectangle $R$.

**Remark 3.10** *The following observations are trivial:*

- *We have $d^{iR} \geq 0$ and $D^{iR} \geq 0$.*

- *For all $x \in R$ we have that $d^{iR} \leq \|x^i - x\| \leq D^{iR}$. Therefore the equation*

$$\min_{x \in R} \phi(\|x^i - x\|) = \min_{r \in [d^{iR}, D^{iR}]} \phi(r)$$

  *is valid for every choice of $\phi$.*

- *The point $x_{max}^{iR}$ is a vertex of $R$ that is not necessarily unique.*

- *If $x^i \in R$, then $d^{iR} = 0$.*

- If $x^i \notin R$, then the computation of $x_{min}^{iR}$ is a convex minimization problem whose solution is on the boundary of the box $R$.

For later considerations we only need the unique distances $d^{iR}$ resp. $D^{iR}$. The points $x_{min}^{iR}$ and $x_{max}^{iR}$ are not needed.

With Definition 3.9 we can specify the area in which the vector $u^n(x) = (\phi(\|x^1 - x\|), \ldots, \phi(\|x^n - x\|))^T$ in (3.1) ranges. This area depends on the minimal respectively maximal distances and the monotonicity of the radial basis function.

**Remark 3.11** *If $\phi$ is monotonically decreasing, then for all $x \in R$ the vector $u^n(x)$ from (3.1) fulfills*

$$u_i^n(x) \in [\phi(D^{iR}), \phi(d^{iR})] \text{ for } i = 1, \ldots, n.$$

*We define the box*

$$V := \big[\phi(D^{1R}), \phi(d^{1R})\big] \times \cdots \times \big[\phi(D^{nR}), \phi(d^{nR})\big]$$

*and conclude that for all $x \in R$ we have*

$$u^n(x) \in V \subset \mathbb{R}_+^n.$$

**Remark 3.12** *If $\phi$ is monotonically increasing, then for all $x \in R$ the components of the vector $u^n(x)$ fulfill*

$$u_i^n(x) \in [\phi(d^{iR}), \phi(D^{iR})] \text{ for } i = 1, \ldots, n.$$

*In this case, we define the box*

$$V := \big[\phi(d^{1R}), \phi(D^{1R})\big] \times \cdots \times \big[\phi(d^{nR}), \phi(D^{nR})\big]$$

*and we get for all $x \in R$*

$$u^n(x) \in V \subset \mathbb{R}_+^n.$$

## 3.2.2 Useful properties of $\phi(r)$ and lower bounds for $s_n(x)$

In the following sections we distinguish between the radial basis functions that are monotone and the thin plate spline which is not. We will see that the formulation of lower bounds for $s_n(x)$ is very similar in both cases.

#### 3.2.2.1 Gaussians, inverse multiquadrics, multiquadrics and cubic spline: Lower bounds for $s_n(x)$

**Lemma 3.13** *The* multiquadrics *($\kappa < 1$) and the* cubic spline *(surface spline with $\kappa = 3$) are strictly monotonically increasing. The* inverse multiquadrics *as well as the* Gaussians *are strictly monotonically decreasing.*

**Proof:** Monotonicity can be verified by checking the sign of the first derivative.

For the *multiquadrics*, i.e. $\phi(r) = (r^2 + w^2)^\kappa$ with $\kappa > 0, \kappa \notin \mathbb{N}, w \in \mathbb{N}_0$, the derivative is as follows

$$\phi'(r) = 2r\kappa(r^2 + w^2)^{\kappa-1} \begin{cases} > 0 & \text{iff } r > 0 \\ = 0 & \text{iff } r = 0. \end{cases}$$

For the *surface spline* with $\kappa = 3$, i.e. $\phi(r) = r^3$, we have the derivative

$$\phi'(r) = 3r^2 \begin{cases} > 0 & \text{iff } r > 0 \\ = 0 & \text{iff } r = 0. \end{cases}$$

For the *inverse multiquadrics*, i.e. $\phi(r) = (r^2 + w^2)^\kappa$ with $\kappa < 0, w \in \mathbb{N}_0$, we have the derivative

$$\phi'(r) = 2r\kappa(r^2 + w^2)^{\kappa-1} \begin{cases} < 0 & \text{iff } r > 0 \\ = 0 & \text{iff } r = 0. \end{cases}$$

And the derivative of the *Gaussians*, i.e. $\phi(r) = \exp(-wr^2)$ with $w \in \mathbb{N}_0$, is

$$\phi'(r) = -2wr \exp(-wr^2) \begin{cases} < 0 & \text{iff } r > 0 \\ = 0 & \text{iff } r = 0. \end{cases}$$

$\square$

The monotonicity and Remarks 3.10, 3.11 and 3.12 enable us to formulate lower bounds for the interpolating functions $s(x)$.

**Corollary 3.14** *For a given interpolating function $s_n(x)$ the following are lower bounds of $s_n(x)$ on $R = [a, b] \subset \mathbb{R}_+^d$, depending on the type of radial basis function used for $s_n(x)$:*

**i)** inverse multiquadric, Gaussians*:*

$$s_n(x) = \sum_{i=1}^n \lambda_i \phi(\|x - x^i\|) = \lambda^T u^n(x)$$

$$\geq \sum_{\lambda_i > 0} \lambda_i \phi(D^{iR}) + \sum_{\lambda_i < 0} \lambda_i \phi(d^{iR})$$

34

**ii)** multiquadrics, surface spline with $\kappa = 1$:

$$s_n(x) = \sum_{i=1}^{n} \lambda_i \phi(\|x - x^i\|) + c = \lambda^T u^n(x) + c$$

$$\geq \sum_{\lambda_i > 0} \lambda_i \phi(d^{iR}) + \sum_{\lambda_i < 0} \lambda_i \phi(D^{iR}) + c$$

**iii)** Surface spline with $\kappa = 3$, i.e. cubic spline:

$$s_n(x) = \sum_{i=1}^{n} \lambda_i \phi(\|x - x^i\|) + c_1 + \sum_{j=2}^{d+1} c_j x_{j-1} = \lambda^T u^n(x) + c_1 + \sum_{j=2}^{d+1} c_j x_{j-1}$$

$$\geq \sum_{\lambda_i > 0} \lambda_i \phi(d^{iR}) + \sum_{\lambda_i < 0} \lambda_i \phi(D^{iR}) + c_1 + \sum_{\substack{c_j < 0 \\ j \neq 1}} c_j b_{j-1} + \sum_{\substack{c_j > 0 \\ j \neq 1}} c_j a_{j-1}.$$

*We will denote all lower bounds by $\Lambda_s$ and keep in mind that the form of $\Lambda_s$ depends on the chosen radial basis function.*

### 3.2.2.2 Thin plate spline: Lower bounds for $s_n(x)$

Now we want to compute lower bounds for $s_n(x)$ in case we are using the *thin plate spline*

$$\phi(r) = r^2 \log r.$$

Note that $\lim_{r \to +0} r^2 \log r = 0$ and therefore we can define $\phi(0) := 0$. In contrast to the other radial basis functions the thin plate spline is neither monotonically increasing nor decreasing. Therefore monotonicity cannot be used like in the other cases. Let us now consider the thin plate spline in detail. The first derivative is

$$\phi'(r) = 2r \log r + r.$$

It is easy to verify that $\lim_{r \to +0} r \log r = 0$ and $\phi'(e^{-\frac{1}{2}}) = 0$. Therefore $r = 0$ and $r = e^{-\frac{1}{2}}$ are potential optima. As the second derivative is

$$\phi''(r) = 2 \log r + 3$$

we have $\lim_{r \to +0} \phi''(r) = -\infty < 0$ and $\phi''(e^{-\frac{1}{2}}) = 2 > 0$. Therefore, $\phi$ has a local maximum at $r = 0$ and a global minimum at $r = e^{-\frac{1}{2}}$. As a result, the optimization of the thin plate spline on an interval $[d, D]$ leads to the following numbers:

$$\min_{r \in [d,D]} \phi(r) = \begin{cases} \phi(D) & = D^2 \log D & \text{if} & D < e^{-\frac{1}{2}} \\ \phi(e^{-\frac{1}{2}}) & = -\frac{1}{2e} & \text{if} & d \leq e^{-\frac{1}{2}} \leq D \\ \phi(d) & = d^2 \log d & \text{if} & d > e^{-\frac{1}{2}} \end{cases}$$

and

$$\max_{r\in[d,D]} \phi(r) = \begin{cases} \phi(d) & = d^2 \log d & \text{if} & D < e^{-\frac{1}{2}} \\ \max\{\phi(d), \phi(D)\} & & \text{if} & d \le e^{-\frac{1}{2}} \le D \\ \phi(D) & = D^2 \log D & \text{if} & d > e^{-\frac{1}{2}}. \end{cases}$$

Here the numbers $d$ and $D$ have not been specified until now. Depending on the considered box $R$ and point $x^i$ we compute the numbers $d^{iR}$ and $D^{iR}$, where $d^{iR}$ resp. $D^{iR}$ denotes the minimum distance respective the maximum distance between $x^i$ and $R$. Furthermore let

$$\phi_{min}^{iR} := \min_{r\in[d^{iR},D^{iR}]} \phi(r)$$

and

$$\phi_{max}^{iR} := \max_{r\in[d^{iR},D^{iR}]} \phi(r)$$

be the solutions of the optimization problems according to the explanations above, i.e. we replace the interval $[d, D]$ by $[d^{iR}, D^{iR}]$ for each point $x^i$ and compute the minimum respectively maximum. Having done this, we are able to formulate a lower bound for $s_n(x)$ as a continuation of Corollary 3.14:

**Corollary 3.15** *For a given interpolating function $s_n(x)$ the following bounds are lower bounds of $s_n(x)$ on $R = [a, b] \subset \mathbb{R}_+^d$:*

**iv)** Thin plate spline:

$$s_n(x) = \sum_{i=1}^{n} \lambda_i \phi(\|x - x^i\|) + c_1 + \sum_{j=2}^{d+1} c_j x_{j-1} = \lambda^T u^n(x) + c_1 + \sum_{j=2}^{d+1} c_j x_{j-1}$$

$$\ge \sum_{\lambda_i>0} \lambda_i \phi_{min}^{iR} + \sum_{\lambda_i<0} \lambda_i \phi_{max}^{iR} + c_1 + \sum_{\substack{c_j<0 \\ j\neq 1}} c_j b_{j-1} + \sum_{\substack{c_j>0 \\ j\neq 1}} c_j a_{j-1}.$$

*We will also denote this lower bound by $\Lambda_s$.*

## 3.3 Useful properties of the matrix $\Phi$ and the (inverse of the) matrix $A$

In order to compute lower bounds for the *auxiliary problem* respectively *weighted auxiliary problem* from Definitions 3.3 and 3.6 let us consider the function

$$z^T A^{-1} z$$

where $z$ lies in an appropriate, but not otherwise specified area. This function coincides up to a constant and a sign with the function in the optimization problem described in Definition 3.3. Therefore it is desirable to compute lower bounds for it. Due to the fact that we solely deal with matrices $A$ defined in Theorem 2.7 we can assume that $A$ is always nonsingular. In the simple case we had $A = \Phi$, which means that the underlying function $\phi$ was (unconditionally) positive definite. The reason for the use of the matrix $P$ (and $P^T$) to expand the matrix $\Phi$ was to keep resp. to restore the nonsingularity of $A$ if $\Phi$ itself was not guaranteed to be nonsingular. For the computation of lower bounds for the problem above it is helpful to know the structure of $A^{-1}$ in detail. In the following sections we will mainly be interested in the properties of the matrix $A$ respectively $A^{-1}$. As an introduction we will start with some specific properties of the matrix $\Phi$ and continue with the general interpolation matrix $A$ respectively $A^{-1}$.

### 3.3.1 The behaviour of $\Phi$

First of all, we will give the formal definition of the so called "distance matrix" $\Phi$, which we have already used in the introduction (see (2.9)).

**Definition 3.16** *Given a radial basis function* $\phi : \mathbb{R}_+ \to \mathbb{R}$ *and a set of points* $\{x^1, \ldots, x^n\}$ *we define the* distance matrix $\Phi$ *as the matrix with the entries*

$$\Phi_{ij} = \phi(\|x^i - x^j\|) \quad \text{for all} \quad i, j \in \{1, \ldots, n\}.$$

The essential feature of $\Phi$, proved for the two radial basis functions *inverse multiquadratics* and the *Gaussians* ([36, 45]), will be cited in Proposition 3.17.

**Proposition 3.17** *For the* inverse multiquadratics *and the* Gaussians *the matrix* $\Phi$ *is positive definite for any choice of pairwise different points* $\{x^1, \ldots, x^n\}$.

**Proof:** For the Gaussian case, see [45, Theorem 6.10]; the proof for the inverse multiquadric follows from [45, Theorem 7.15]. □

Due to Definition 2.2 we have that Proposition 3.17 is a corollary of Theorem 2.4 at the same time.

As a consequence of Proposition 3.17, the inverse matrix $\Phi^{-1}$ is positive definite as well. As a consequence we have Corollary 3.18.

**Corollary 3.18** *For the* inverse multiquadratics *and the* Gaussians *the optimization problem*

$$\min z^T \Phi^{-1} z \quad s.t. \ z \in [a, b] \subset \mathbb{R}_+^d$$

*is a convex optimization problem whose solution is on the boundary of* $[a, b]$.

The positive definiteness of $\Phi$ does not hold in case of the other radial basis functions. Therefore a result like in Corollary 3.18 is not available. However, a similar but weaker statement compared to Proposition 3.17 can be deduced from Theorem 2.4 for the *multiquadrics* and the *linear spline* (surface spline with $\kappa = 1$).

**Corollary 3.19** *For the* multiquadrics *with* $\kappa \in (0, 1)$ *and the* linear spline *we have that*

$$-\lambda^T \Phi \lambda > 0$$

*for any choice of pairwise different points* $\{x^1, \ldots, x^n\}$ *and any nonzero vector* $\lambda = (\lambda_1, \ldots, \lambda_n)^T$ *with* $\sum_{i=1}^n \lambda_i = 0$.

In contrast to Proposition 3.17 we no longer have positive definiteness in case of *multiquadrics* and the *linear spline*. Instead, Corollary 3.19 describes a kind of positive definiteness of the matrix $-\Phi$ over a specified set of $\lambda$'s. This feature is available similarly for the other radial basis functions *thin plate spline* and *cubic spline* as well.

**Corollary 3.20** *For the* thin plate spline *and the* cubic spline *(surface splines with* $\kappa = 2, 3$*) we have that*

$$\lambda^T \Phi \lambda > 0$$

*for any choice of pairwise different points* $\{x^1, \ldots, x^n\}$ *and any nonzero vector* $\lambda = (\lambda_1, \ldots, \lambda_n)^T$ *with*

$$\sum_{i=1}^n \lambda_i = \sum_{i=1}^n \lambda_i x_1^i = \sum_{i=1}^n \lambda_i x_2^i = \ldots = \sum_{i=1}^n \lambda_i x_d^i = 0. \tag{3.4}$$

**Proof:**

We have that the monomials $1, x_1, \ldots, x_d$ are a basis of $\Pi_1^d$. Therefore the conditions (3.4) are equivalent to

$$\sum_{i=1}^{n} \lambda_i q(x^i) = 0 \text{ for all } q \in \Pi_1^d.$$

With Definition 2.2 and Theorem 2.4 we have the desired result. $\qquad\square$

If we compare conditions (3.4) with Remark 3.1 iii) we see that we can express (3.4) equivalently by

$$P^T \lambda = 0.$$

That is, the property of a function $\phi$ being conditionally positive definite introduces directly a property of the involved matrix $\Phi$, namely that $\Phi$ *is strictly copositive with respect to* $\ker(P^T)$. Similarly we have that for the *multiquadrics* with $\kappa \in (0,1)$ and the *linear spline* the matrix $-\Phi$ is strictly copositive with respect to $\ker(P^T)$. We will use this in Section 3.3.2.

## 3.3.2 Theoretical aspects concerning $A \neq \Phi$

Now we consider the general case with

$$A = \begin{pmatrix} \Phi & P \\ P^T & 0 \end{pmatrix}.$$

We observe that the expanded matrix $A$ still contains the matrix $\Phi$ as a block and, in addition, the matrix $\Phi$ respectively $(-\Phi)$ is still copositive with respect to $\ker(P^T)$ due to the conditional positive definiteness of the function $\phi$ respectively $-\phi$. Therefore it is desirable to use this copositivity for computing lower bounds. This requires an explicit form of the inverse of $A$, that is, a description of $A^{-1}$ as a block matrix including the original matrix $\Phi$. Indeed there exist such formulas for the inverse of $2 \times 2$-block matrices at most one of whose diagonal blocks is singular (see e.g. [34]). Unfortunately the blockmatrices on the diagonal of $A$ are potentially both singular in our case. Therefore it requires some effort to find an appropriate formula for $A^{-1}$. Before we formulate this we provide the necessary tools.

**Definition 3.21** *For a real matrix $M$ the* Moore-Penrose pseudoinverse *is the unique matrix $M^\dagger$ satisfying all of the following conditions:*

$$MM^\dagger M = M, \qquad\qquad M^\dagger M M^\dagger = M^\dagger,$$
$$(MM^\dagger)^T = MM^\dagger, \qquad\qquad (M^\dagger M)^T = M^\dagger M.$$

In addition we have that

$$(M^T)^\dagger = (M^\dagger)^T. \tag{3.5}$$

Now so called *projection matrices*, deduced from the Moore-Penrose pseudoinverse have well known properties specified in Remark 3.22:

**Remark 3.22** *We have that*

a) $(MM^\dagger)^2 = MM^\dagger$,

b) $MM^\dagger$ *resp.* $(I - MM^\dagger)$ *are orthogonal projection operators onto* $\mathrm{Im}(M)$ *resp.* $\ker(M^T)$,

c) $M^\dagger M$ *resp.* $(I - M^\dagger M)$ *are orthogonal projection operators onto* $\mathrm{Im}(M^T)$ *resp.* $\ker(M)$,

d) *if the rows of* $M$ *are linearly independent, then* $MM^\dagger = I$.

Next we define a matrix that will play a crucial role in the remainder. Let $P$ and $\Phi$ the matrices already introduced. Then we define the matrix $Q$ as

$$Q := (I - PP^\dagger)\Phi(I - PP^\dagger).$$

Lemma 3.23 outlines some properties of $Q$ and $Q^\dagger$, respectively.

**Lemma 3.23** *We have that*

a) $Q$ *and* $Q^\dagger$ *are symmetric.*

b) *If* $\Phi$ *is copositive with respect to* $\ker(P^T)$, *then* $Q$ *and* $Q^\dagger$ *are positive semidefinite.*

**Proof:**

a) We have that

$$\begin{aligned}
Q^T &= [(I - PP^\dagger)\Phi(I - PP^\dagger)]^T \\
&= (I - PP^\dagger)^T \Phi^T (I - PP^\dagger)^T \\
&= (I - PP^\dagger)\Phi(I - PP^\dagger) \\
&= Q,
\end{aligned}$$

where the symmetry of $(I - PP^\dagger)$ follows from Definition 3.21. Moreover, due to the symmetry of $Q$ and (3.5) we have that $(Q^\dagger)^T = (Q^T)^\dagger = Q^\dagger$.

b) Let $x \in \mathbb{R}^n$. Then we have that $(I - PP^\dagger)x \in \ker(P^T)$. Due to the copositivity of $\Phi$ over $\ker(P^T)$ we get

$$x^T Q x = x^T (I - PP^\dagger)\Phi(I - PP^\dagger)x \geq 0 \quad \text{for all } x \in \mathbb{R}^n.$$

Therefore, $Q$ is positive semidefinite. This implies that

$$x^T Q^\dagger x = x^T Q^\dagger Q Q^\dagger x = (Q^\dagger x)^T Q (Q^\dagger x) \geq 0 \quad \text{for all } x \in \mathbb{R}^n.$$

Therefore, $Q^\dagger$ is positive semidefinite as well. $\qquad\square$

Lemmas 3.24 and 3.25 provide a deeper insight into the behaviour of $Q$. We will see that these features are essential for the proof of Theorem 3.27.

**Lemma 3.24** *Let $\Phi$ and $P$ be matrices as defined in the previous chapters. Then*

$$Q^\dagger Q x = Q Q^\dagger x = x \quad \text{for all } x \in \ker(P^T)$$

**Proof:** We prove the assertion with the following two steps a) and b):

**Part a):**
We show that the map $x \mapsto Qx = (I - PP^\dagger)\Phi(I - PP^\dagger)x$ is injective on $\ker(P^T)$. To see this we assume that there exists an $x \in \ker(P^T)$ with $x \neq 0$ and $Qx = 0$. Due to $(I - PP^\dagger)x = x$ for all $x \in \ker(P^T)$ we conclude that

$$x^T \Phi x = ((I - PP^\dagger)x)^T \Phi(I - PP^\dagger)x = x^T Q x = x^T 0 = 0$$

which is a contradiction to the strict copositivity of $\Phi$ with respect to $\ker(P^T)$. Therefore the only $x \in \ker(P^T)$ with $Qx = 0$ is $x = 0$ which proves injectivity of the map $x \mapsto Qx$ on $\ker(P^T)$.

Since for linear maps injectivity is equivalent to bijectivity, we have that the map

$$x \mapsto Qx$$

is bijective on $\ker(P^T)$.

**Part b):**
Observe that $\text{Im}(Q) \subseteq \ker(P^T)$, since for any $x \in \mathbb{R}^n$ we have that

$$Qx = (I - PP^\dagger)\underbrace{\Phi(I - PP^\dagger)x}_{=:z'} = (I - PP^\dagger)z' \in \ker(P^T),$$

because $(I - PP^\dagger)$ is the orthogonal projection operator onto $\ker(P^T)$.

Now let $x \in \ker(P^T)$. From Remark 3.22 c) and the symmetry of $Q$ we know that

$$z := Q^\dagger Q x \in \mathrm{Im}(Q) \subseteq \ker(P^T)$$

Moreover we have that

$$Qz = QQ^\dagger Q x = Qx.$$

Therefore, we must have $x = z$ by part a), or equivalently

$$x = Q^\dagger Q x.$$

Since $x \in \ker(P^T)$ was arbitrary, this proves the statement. □

With Lemma 3.24 we can prove the Lemma 3.25:

**Lemma 3.25** *We have that*

$$(I - PP^\dagger) = QQ^\dagger = Q^\dagger Q.$$

**Proof:**

We only have to show the first equality, the second follows from symmetry of $Q$ and $Q^\dagger$. First we prove that for any $x \in \mathbb{R}^n$ the vector $(QQ^\dagger x - x)$ and $\ker(P^T)$ are orthogonal: Let $y \in \ker(P^T)$ and let $x \in \mathbb{R}^n$. Using $(QQ^\dagger)^T = QQ^\dagger$ and $QQ^\dagger y = y$, we get:

$$\begin{aligned}
y^T(QQ^\dagger x - x) &= y^T QQ^\dagger x - y^T x \\
&= ((QQ^\dagger)^T y)^T x - y^T x \\
&= (QQ^\dagger y)^T x - y^T x \\
&= y^T x - y^T x \\
&= 0.
\end{aligned}$$

Consequently, the orthogonal projection of $QQ^\dagger x$ onto $\ker(P^T)$ equals the orthogonal projection of $x$ onto $\ker(P^T)$, i.e.

$$(I - PP^\dagger)QQ^\dagger x = (I - PP^\dagger)x \quad \text{for all } x \in \mathbb{R}^n. \tag{3.6}$$

From $(I - PP^\dagger)^2 = (I - PP^\dagger)$ it follows that

$$\begin{aligned}
(I - PP^\dagger)Q &= (I - PP^\dagger)(I - PP^\dagger)\Phi(I - PP^\dagger) \\
&= (I - PP^\dagger)\Phi(I - PP^\dagger) \\
&= Q,
\end{aligned}$$

and therefore we conclude with (3.6) that

$$QQ^\dagger = (I - PP^\dagger).$$

□

**Lemma 3.26** *We have that*

$$(I - PP^\dagger)Q^\dagger(I - PP^\dagger) = Q^\dagger.$$

**Proof:**

Using Lemma 3.25 twice, we get

$$(I - PP^\dagger)Q^\dagger(I - PP^\dagger) = Q^\dagger QQ^\dagger QQ^\dagger = Q^\dagger QQ^\dagger = Q^\dagger.$$

$\square$

With these lemmas we can now characterize the inverse of $A$:

**Theorem 3.27** *The inverse of* $A = \begin{pmatrix} \Phi & P \\ P^T & 0 \end{pmatrix}$ *is the matrix* $A^{-1} = \begin{pmatrix} Q^\dagger & B \\ B^T & C \end{pmatrix}$
*with*

$$B = (P^\dagger)^T - Q^\dagger\Phi(P^\dagger)^T$$
$$C = P^\dagger\left[\Phi Q^\dagger\Phi - \Phi\right](P^\dagger)^T$$

*where, as before,* $Q = (I - PP^\dagger)\Phi(I - PP^\dagger)$.

**Proof:** We have to prove that

$$\begin{pmatrix} \Phi & P \\ P^T & 0 \end{pmatrix}\begin{pmatrix} Q^\dagger & B \\ B^T & C \end{pmatrix} =: \begin{pmatrix} D_1 & D_2 \\ D_3 & D_4 \end{pmatrix} = \begin{pmatrix} I & 0 \\ 0 & I \end{pmatrix}.$$

First we show that $D_1 = I$. We have:

$$
\begin{aligned}
D_1 &= \Phi Q^\dagger + PB^T \\
&= \Phi Q^\dagger + P[(P^\dagger)^T - Q^\dagger\Phi(P^\dagger)^T]^T \\
&= \Phi Q^\dagger + P[P^\dagger - P^\dagger\Phi Q^\dagger] \\
&= \Phi Q^\dagger + PP^\dagger - PP^\dagger\Phi Q^\dagger \\
&= (I - PP^\dagger)\Phi Q^\dagger + PP^\dagger \qquad\qquad\qquad (3.7) \\
&= (I - PP^\dagger)\Phi(I - PP^\dagger)Q^\dagger(I - PP^\dagger) + PP^\dagger \\
&= QQ^\dagger(I - PP^\dagger) + PP^\dagger \\
&= (I - PP^\dagger)(I - PP^\dagger) + PP^\dagger \\
&= I - 2PP^\dagger + (PP^\dagger)^2 + PP^\dagger \\
&= I,
\end{aligned}
$$

43

where in lines $6, 8$ and $9$ we use Lemmas 3.26 and 3.25 and Remark 3.22 a), respectively.

Next, we prove that $D_2 = 0$:

$$
\begin{aligned}
D_2 &= \Phi B + PC \\
&= \Phi[(P^\dagger)^T - Q^\dagger \Phi (P^\dagger)^T] + P[P^\dagger (\Phi Q^\dagger \Phi - \Phi)(P^\dagger)^T] \\
&= \Phi (P^\dagger)^T - \Phi Q^\dagger \Phi (P^\dagger)^T + PP^\dagger (\Phi Q^\dagger \Phi - \Phi)(P^\dagger)^T \\
&= \Phi (P^\dagger)^T - PP^\dagger \Phi (P^\dagger)^T - \Phi Q^\dagger \Phi (P^\dagger)^T + PP^\dagger \Phi Q^\dagger \Phi (P^\dagger)^T \\
&= (I - PP^\dagger)\Phi (P^\dagger)^T - \underbrace{(I - PP^\dagger)\Phi Q^\dagger}_{=D_1 - PP^\dagger = I - PP^\dagger, \text{ see } (3.7)} \Phi (P^\dagger)^T \\
&= (I - PP^\dagger)\Phi (P^\dagger)^T - (I - PP^\dagger)\Phi (P^\dagger)^T \\
&= 0
\end{aligned}
$$

It is not difficult to see that $D_3 = 0$:

$$
\begin{aligned}
D_3 &= P^T Q^\dagger \\
&= P^T (I - PP^\dagger) Q^\dagger (I - PP^\dagger).
\end{aligned}
$$

Due to Remark 3.22 b) the matrix $(I - PP^\dagger)$ is the orthogonal projection onto $\ker(P^T)$. Therefore all columns of $(I - PP^\dagger)Q^\dagger(I - PP^\dagger)$ are elements of $\ker(P^T)$ and consequently $D_3 = 0$.

Finally, we argue that $D_4 = I$:

$$
\begin{aligned}
D_4 &= P^T B \\
&= P^T[(P^\dagger)^T - Q^\dagger \Phi (P^\dagger)^T] \\
&= P^T (P^\dagger)^T - P^T Q^\dagger \Phi (P^\dagger)^T \\
&= P^T (P^\dagger)^T - P^T (I - PP^\dagger)Q^\dagger (I - PP^\dagger)\Phi (P^\dagger)^T.
\end{aligned}
$$

Since the rows of $P^T$ are linearly independent, we get from Remark 3.22 d) that $P^T(P^\dagger)^T = I$. The term $P^T(I - PP^\dagger)Q^\dagger(I - PP^\dagger)\Phi(P^\dagger)^T$ equals 0 by the same arguments as above. Hence, we get that $D_4 = I$. $\qquad\square$

The structure of the inverse in Theorem 3.27 fits to the applications we have in mind. A crucial result is the positive semidifineteness of $Q^\dagger$, which enables us to compute the lower bounds we present afterwards. We point out that there exist similar formulas [10, 15, 33] for the inverse we presented in Theorem 3.27 . In [10] the requirements for the matrix $A$, in detail for the matrices $\Phi$ and $P$, are much more general and therefore the resulting block within the inverse corresponding to

$Q^\dagger$ is not guaranteed having special properties. In [15] the requirements for the matrix $A$ are similar to the requirements in Theorem 3.27: The block matrix $\Phi$ has to be nonnegative definite and may be singular. In contrast the rank of $P$ may be deficient in rank. With these requirements the authors in [15] present a formula for a *generalized inverse*, i.e. a matrix, which only has to fulfill the first of the four conditions presented in Definition 3.21. The formula in [15] is a generalization of the inverse presented in Theorem 3.27. This can be shown by using Lemma 3.26 and by using the appropriate Moore-Penrose-Inverses everywhere in [15, Theorem 3.1] where a generalized inverse is required. In [33] the results of [15] are extended for a more general matrix $A$.

We finish this section with an additional lemma concerning a connection between the matrices $Q^\dagger$ and $P$.

**Lemma 3.28** *We have that*

$$\ker(Q^\dagger) = \operatorname{Im}(P).$$

**Proof:**
It is well known (see, e.g., [23]) that the solution set (if nonempty) of a system $Ax = b$ is given by $\left\{ A^\dagger b + (I - A^\dagger A)y \mid y \in \mathbb{R}^n \right\}$. Applying this to the system

$$Q^\dagger u = 0$$

gives

$$\ker(Q^\dagger) = \left\{ (I - QQ^\dagger)y \mid y \in \mathbb{R}^n \right\} = \operatorname{Im}(I - QQ^\dagger).$$

From Lemma 3.25 we have that

$$I - QQ^\dagger = I - (I - PP^\dagger) = PP^\dagger.$$

Therefore we have

$$\ker(Q^\dagger) = \operatorname{Im}(I - QQ^\dagger) = \operatorname{Im}(PP^\dagger) = \operatorname{Im}(P),$$

where the last equality comes from Remark 3.22. $\qquad\square$

## 3.4 Computation of lower bounds for the auxiliary problem and the weighted auxiliary problem

We have seen that the interpolation matrices that occur within the *auxiliary problem* respectively *weighted auxiliary problem* differ depending on the chosen radial basis function. In the case of *Gaussians* and *inverse multiquadrics* the inverse of the distance matrix $\Phi$ is needed. In all the other cases the inverse of the expanded matrix $A$ is needed. In the following, we will distinguish between both scenarios.

### 3.4.1 Lower bounds if $A = \Phi$

In Section 3.2 we used the positive definiteness of the *inverse multiquadrics* and *Gaussians* to compute lower bounds for the interpolating function $s_n(x)$. For the *auxiliary problem* resp. *weighted auxiliary problem* in Definitions 3.3 resp. 3.6 the calculation of a lower bound needs more effort. Due to the fact that the objective of these problems in the case of $A = \Phi$ mainly consists (see Remark 3.1 part i) of the squared form

$$u^n(x)^T \Phi^{-1} u^n(x),$$

where

$$u^n(x) = (\phi(\|x^1 - x\|), \ldots, \phi(\|x^n - x\|)),$$

we are interested in the behaviour of the product of functions like

$$\phi(\|x^i - x\|)\phi(\|x^j - x\|) \text{ with } i, j \in \{1, \ldots, n\}.$$

More general, because of Remarks 3.11 and 3.12, we are interested in the product that is definded in Definition 3.29.

**Definition 3.29** *Let $\phi$ be one of the functions* inverse multiquadrics *or* Gaussians *and let $R \subset \mathbb{R}^d$ be a box. Let $x^i$ and $x^j$ be any two points in $\mathbb{R}^d$ with the corresponding $d^{iR}$ resp. $D^{iR}$ and $d^{jR}$ resp. $D^{jR}$ according to Definition 3.9. For $u \in \left[d^{iR}, D^{iR}\right]$ and $v \in \left[d^{jR}, D^{jR}\right]$, we define*

$$\Psi(u, v) := \phi(u)\phi(v) \quad on \quad \left[d^{iR}, D^{iR}\right] \times \left[d^{jR}, D^{jR}\right] \subset \mathbb{R}_+^2$$

*as the* product of basisfunctions.

**Proposition 3.30** *Let* $\phi$ *be either the* (i) *inverse multiquadrics or the* (ii) *Gaussians. Then any vector* $p \in \mathbb{R}^{>0} \times \mathbb{R}^{>0}$ *is a descent direction for* $\Psi$ *in any point* $(u, v) \neq (0, 0)$.

**Proof:** Let $(u, v) \neq (0, 0)$ and let $p \in \mathbb{R}^{>0} \times \mathbb{R}^{>0}$, that is, let $p = (p_1, p_2)$ with $p_1 > 0$ and $p_2 > 0$. We show that the directional derivative $\nabla \Psi(u, v)^T \cdot p < 0$.

**(i)** inverse multiquadrics:

The gradient of

$$\Psi(u, v) = (u^2 + w^2)^{\kappa}(v^2 + w^2)^{\kappa} = \left[(u^2 + w^2)(v^2 + w^2)\right]^{\kappa}$$

is

$$\nabla \Psi(u, v) = \kappa \left[(u^2 + w^2)(v^2 + w^2)\right]^{\kappa - 1} \left((v^2 + w^2)2u, (u^2 + w^2)2v\right)^T.$$

Then the directional derivative is

$$\nabla \Psi(u, v)^T \cdot p = \underbrace{\kappa \left[(u^2 + w^2)(v^2 + w^2)\right]^{\kappa - 1}}_{<0} \underbrace{\left((v^2 + w^2)2up_1 + (u^2 + w^2)2vp_2\right)}_{>0}$$

$$< 0.$$

**(ii)** Gaussians:

The gradient of

$$\Psi(u, v) = \exp\left(-wu^2\right)\exp\left(-wv^2\right)$$

is

$$\nabla \Psi(u, v) = -2\exp\left(-wu^2\right)\exp\left(-wv^2\right)(u, v)^T.$$

Then the directional derivative is

$$\nabla \Psi(u, v)^T p = \underbrace{-2\exp\left(-wu^2\right)\exp\left(-wv^2\right)}_{<0}\underbrace{\left(up_1 + vp_2\right)}_{>0}$$

$$< 0.$$

$\square$

**Proposition 3.31** *Let* $\phi$ *be either the* inverse multiquadrics *or the* Gaussians. *Then* (i) *the minimum of* $\Psi$ *over* $[d^{iR}, D^{iR}] \times [d^{jR}, D^{jR}]$ *is attained in the point* $(D^{iR}, D^{jR})^T$ *and* (ii) *the maximum of* $\Psi$ *over* $[d^{iR}, D^{iR}] \times [d^{jR}, D^{jR}]$ *is attained in* $(d^{iR}, d^{jR})^T$.

**Proof:** Let $[a, b] \subset \mathbb{R}^2_+$ be a nonnegative rectangle. Then we conclude the following:

**(i)** Assume that $t \in \operatorname{int}[a, b]$. Then there obviously exists a point $p \in \mathbb{R}^{>0} \times \mathbb{R}^{>0}$ with

$$t + p = b.$$

Due to Proposition 3.30 the directional derivative $\nabla\Psi(t + s \cdot p)^T p$ is negative for every $s \in (0, 1)$. Thus, and because $\Psi$ is continuous, we have $\Psi(t) > \Psi(b)$ and therefore the minimum is $b$.

**(ii)** Assume that $t \in \operatorname{int}[a, b]$. Then there obviously exists a point $p \in \mathbb{R}^{>0} \times \mathbb{R}^{>0}$ with

$$t - p = a.$$

Due to Proposition 3.30 the directional derivative $\nabla\Psi(t - s \cdot p)^T(-p)$ is positive for every $s \in (0, 1)$. Thus, and because $\Psi$ is continuous, we have $\Psi(t) < \Psi(a)$ and therefore the maximum is $a$.

With $a := (d^{iR}, d^{jR})^T$ and $b := (D^{iR}, D^{jR})^T$ the assertion is proved. $\qquad\square$

As a result, it is possible to formulate the two different lower bounds $\Lambda_a^1$ and $\Lambda_a^2$ for the *auxiliary problem* resp. *weighted auxiliary problem*:

**Corollary 3.32** *Let $\phi$ be an* inverse multiquadric *or a* Gaussian *and let $R = \left[x^L, x^U\right] \subset \mathbb{R}_+^d$ an arbitrary box. Then for the* auxiliary problem *resp.* weighted auxiliary problem *in Definition 3.3 resp. 3.6 and for every $x \in R$ we have:*

**1)** *Consider the objective function*

$$a_n(x) := z^n(x)^T \Phi^{-1} z^n(x) - \phi(0)$$

*of the* auxiliary problem. *We have:*

$$a_n(x) \geq \Lambda_a^1 \quad \text{and} \quad a_n(x) \geq \Lambda_a^2 \quad \text{for all } x \in R$$

*where*

$$\Lambda_a^1 = \sum_{\Phi_{ij}^{-1} > 0} \Phi_{ij}^{-1} \Psi((D^{iR}, D^{jR})^T) + \sum_{\Phi_{ij}^{-1} < 0} \Phi_{ij}^{-1} \Psi((d^{iR}, d^{jR})^T) - \phi(0)$$

*and*

$$\Lambda_a^2 = \min_{u \in V} u^T \Phi^{-1} u - \phi(0),$$

*where the last expression is a convex optimization problem (see Corollary 3.18). The set*

$$V = \left[\phi(D^{1R}), \phi(d^{1R})\right] \times \cdots \times \left[\phi(D^{nR}), \phi(d^{nR})\right]$$

*is taken from Remark 3.11 due to the monotonicity of the* inverse multiquadric *and* Gaussians.

**2)** *Consider the objective function*

$$w_n(x) := \frac{1}{[s_n(x) - f^*]^2} \left( z^n(x)^T \Phi^{-1} z^n(x) - \phi(0) \right)$$

*of the weighted auxiliary problem. Let $\Lambda_a$ be one of the two lower bounds $\Lambda_a^1$ or $\Lambda_a^2$ given in part 1). We have:*

$$w_n(x) \geq \Lambda_w \quad \text{for all} \quad x \in R$$

*where*

$$\Lambda_w = \frac{1}{[\Lambda_s - f^*]^2} \cdot \Lambda_a,$$

*where $\Lambda_s$ is taken from Corollary 3.14.*

**Proof:**

**1)** For the *auxiliary problem*, we have:

$$
\begin{aligned}
a_n(x) &:= z^n(x)^T \Phi^{-1} z^n(x) - \phi(0) \\
&= \sum_{i=1}^{n} \sum_{j=1}^{n} \Phi_{ij}^{-1} \phi(\underbrace{\|x^i - x\|}_{=:u}) \phi(\underbrace{\|x^j - x\|}_{=:v}) - \phi(0) \\
&= \sum_{i=1}^{n} \sum_{j=1}^{n} \Phi_{ij}^{-1} \phi(u) \phi(v) - \phi(0) \\
&= \sum_{i=1}^{n} \sum_{j=1}^{n} \Phi_{ij}^{-1} \Psi(u, v) - \phi(0) \\
&\geq \sum_{\Phi_{ij}^{-1} > 0} \Phi_{ij}^{-1} \Psi((D^{iR}, D^{jR})^T) + \sum_{\Phi_{ij}^{-1} < 0} \Phi_{ij}^{-1} \Psi((d^{iR}, d^{jR})^T) - \phi(0) =: \Lambda_a^1,
\end{aligned}
$$

where the inequality comes from the fact that $\Psi(u, v) > 0$ and

$$(u, v) \in \left[ d^{iR}, D^{iR} \right] \times \left[ d^{jR}, D^{jR} \right].$$

Furthermore, due to Proposition 3.17, we have that $\Phi$ and therefore $\Phi^{-1}$ is positive definite. In the case of *Gaussian* and *multiquadric* we have $z^n(x) = u^n(x)$, and therefore

$$
\begin{aligned}
a_n(x) &:= z^n(x)^T \Phi^{-1} z^n(x) - \phi(0) \\
&= u^n(x)^T \Phi^{-1} u^n(x) - \phi(0) \\
&\geq \min_{u \in V} u^T \Phi^{-1} u - \phi(0).
\end{aligned}
\tag{3.8}
$$

The last inequality holds due to Remarks 3.11 and 3.12.

**2)** Let $\Lambda_a$ be $\Lambda_a^1$ or $\Lambda_a^2$. We start with the facts that

$$s_n(x) \geq \Lambda_s \text{ for all } x \in R$$

and (compare with Algorithm 2.1, step 2 a))

$$\Lambda_s \geq f^*.$$

By using elementarly mathematical operations we conclude

$$s_n(x) - f^* \geq \Lambda_s - f^*$$
$$\Rightarrow \quad (s_n(x) - f^*)^2 \geq (\Lambda_s - f^*)^2$$
$$\Rightarrow \quad \frac{1}{(s_n(x) - f^*)^2} \leq \frac{1}{(\Lambda_s - f^*)^2}.$$

With Lemma 3.4 and Definition 3.6 we finally have the relation

$$w_n(x) = \frac{a_n(x)}{(s_n(x) - f^*)^2} \geq \frac{a_n(x)}{(\Lambda_s - f^*)^2} \geq \frac{\Lambda_a}{(\Lambda_s - f^*)^2}.$$

$$\square$$

## 3.4.2 Lower bounds if $A \neq \Phi$

For computing lower bounds for the *auxiliary problem* and the *weighted auxiliary problem* in case $A \neq \Phi$ we need to do some preliminary work that is to recall some facts as well as to consider a subproblem that will occur later. After doing this in Section 3.4.2.1 we will go further on with the computation of lower bounds in Section 3.4.2.2.

### 3.4.2.1 Preliminaries

We are given a box $R = [x^L, x^U] \subset \mathbb{R}_+^d$, the radial basis function $\phi(r)$ and $n$ interpolation points $x^1, \ldots, x^n$.

Furthermore for $x \in R$ we have that $\|x^i - x\| \in [d^{iR}, D^{iR}]$ describes the distance between the interpolation point $x^i$ and $x$. In addition $d^{iR}$ and $D^{iR}$ is the minimum distance between $x^i$ and $R$ and the maximum distance between $x^i$ and $R$, respectively.

We know that

$$\phi \text{ monotonically} \begin{cases} \text{decreasing} & \Rightarrow \phi(\|x^i - x\|) \in [\phi(D^{iR}), \phi(d^{iR})] \\ \text{increasing} & \Rightarrow \phi(\|x^i - x\|) \in [\phi(d^{iR}), \phi(D^{iR})] \end{cases}$$

For later considerations we need bounds for the expressions

$$\phi(\|x^i - x\|)x_j \quad \text{and} \quad x_i x_j \quad \text{with} \quad x = (x_1, \ldots, x_d) \in R.$$

To keep the effort small, we will only consider the case with $\phi$ monotonically increasing; with $\phi$ monotonically decreasing the treatment is similar.

The estimates of the expressions are as follows:

Due to $x_j \leq x_j^U$ and $\phi(\|x^i - x\|) \geq 0$ we have $\phi(\|x^i - x\|)x_j \leq \phi(\|x^i - x\|)x_j^U$ and overall (note that $R \subset \mathbb{R}_+^d$)

$$\phi(d^{ik})x_j^L \leq \phi(\|x^i - x\|)x_j \leq \phi(D^{ik})x_j^U.$$

Furthermore we have

$$x_i^L x_j^L \leq x_i x_j \leq x_i^U x_j^U.$$

**Theorem 3.33** *Let us consider the optimization problem (with $V$, $P$ and $Q^\dagger$ as usual)*

$$\min_{u \in V} u^T Q^\dagger u. \tag{3.9}$$

*We define*

$$\Lambda_a' := \min_{v \in Proj_{\ker(PT)}(V)} v^T Q^\dagger v.$$

*The number $\Lambda_a'$ is the minimum of (3.9) and therefore a valid lower bound for (3.9).*

**Proof:**

From Lemma 3.23 we know that $Q^\dagger$ is positive semidefinite. Therefore, (3.9) is a convex optimization problem with $u^T Q^\dagger u = 0$ for all $u \in \ker(Q^\dagger)$. Furthermore, by using the Moore-Penrose property $Q^\dagger = Q^\dagger Q Q^\dagger$ twice in the first equation, we have

$$\begin{aligned} \min_{u \in V} u^T Q^\dagger u &= \min_{u \in V} u^T Q^\dagger Q Q^\dagger Q Q^\dagger u \\ &= \min_{u \in V} (Q^\dagger Q u)^T Q^\dagger Q Q^\dagger u \\ &= \min_{u \in V} \underbrace{(Q Q^\dagger u)}_{=:v}^T Q^\dagger (Q Q^\dagger u) \\ &= \min_{v \in V'} v^T Q^\dagger v \end{aligned}$$

51

where $v \in \ker(P^T)$ and $V' = Proj_{\ker(P^T)}(V)$ due to the fact that $QQ^\dagger$ is the orthogonal projection map onto $\ker(P^T)$. □

As a direct consequence of the positive semidefiniteness of $Q^\dagger$ we have that

$$\Lambda_a' = \begin{cases} 0 & \text{if } \ker(Q^\dagger) \cap V \neq \emptyset \\ \min_{v \in Proj_{\ker(P^T)}(V)} v^T Q^\dagger v & \text{if } \ker(Q^\dagger) \cap V = \emptyset. \end{cases}$$

### 3.4.2.2   Computing lower bounds

As a result of Theorem 3.27 the matrix $A^{-1}$ that occurs in the *auxiliary problem* and the *weighted auxiliary problem* has the form

$$A^{-1} = \begin{pmatrix} Q^\dagger & B \\ B^T & C \end{pmatrix},$$

with $Q^\dagger$ positive semidefinite. We use this fact for computing lower bounds as described in the next corollaries.

Note that in case of *multiquadric* and *linear spline* the matrix $B$ is a column vector and $C$ is a $1 \times 1$ matrix.

**Corollary 3.34** *Let $\phi(r)$ be the* multiquadric *or the* linear spline *(both monotonically increasing) and let $R = \left[ x^L, x^U \right] \subset \mathbb{R}_+^d$ an arbitrary box. Then for the auxiliary problem resp. weighted auxiliary problem in Definitions 3.3 resp. 3.6 and for every $x \in R$ we have:*

1) *For the objective function*

$$a_n(x) := \phi(0) - z^n(x)^T A^{-1} z^n(x)$$

*of the auxiliary problem, we have*

$$a_n(x) \geq \Lambda_a \quad \text{for all} \quad x \in R$$

*where*

$$\Lambda_a = \phi(0) + \Lambda_a' + 2 \sum_{B_i > 0} B_i \phi(d^{iR}) + 2 \sum_{B_i < 0} B_i \phi(D^{iR}) + C$$

*and $\Lambda_a'$ is the number from Theorem 3.33.*

52

**2)** *Consider the objective function*

$$w_n(x) := \frac{1}{[s_n(x) - f^*]^2} \left( \phi(0) - z^n(x)^T A^{-1} z^n(x) \right)$$

*of the weighted auxiliary problem. We have:*

$$w_n(x) \geq \Lambda_w \quad \text{for all} \quad x \in R$$

*where*

$$\Lambda_w = \frac{1}{[\Lambda_s - f^*]^2} \cdot \Lambda_a.$$

**Proof:**

**1)** Let $x \in R$. Recall that $z^n(x) = (u^n(x)^T, 1)^T$ and $u^n(x) = (\phi(\|x^1 - x\|), \ldots, \phi(\|x^n - x\|))^T$ (cf. Section 3.1). Furthermore recall that for the matrix

$$-A = \begin{pmatrix} -\Phi & -P \\ -P^T & 0 \end{pmatrix}$$

the matrix $-\Phi$ is, due to the conditional positive definiteness of $-\phi$, copositive with respect to $\ker(P^T)$. With Lemma 3.23 we conclude, that the corresponding block in $(-A)^{-1}$, called $Q^\dagger$, is positive semidefinite. Using that $-A^{-1} = (-A)^{-1}$ we have for the *auxiliary problem*:

$$
\begin{aligned}
a_n(x) &:= \phi(0) - z^n(x)^T A^{-1} z^n(x) \\
&= \phi(0) + z^n(x)^T (-A)^{-1} z^n(x) \\
&= \phi(0) + \begin{pmatrix} u^n(x)^T & 1 \end{pmatrix} \begin{pmatrix} Q^\dagger & B \\ B^T & C \end{pmatrix} \begin{pmatrix} u^n(x) \\ 1 \end{pmatrix} \\
&= \phi(0) + \begin{pmatrix} u^n(x)^T Q^\dagger + B^T & , & u^n(x)^T B + C \end{pmatrix} \begin{pmatrix} u^n(x) \\ 1 \end{pmatrix} \\
&= \phi(0) + \left( u^n(x)^T Q^\dagger u^n(x) + B^T u^n(x) + u^n(x)^T B + C \right) \\
&= \phi(0) + u^n(x)^T Q^\dagger u^n(x) + 2 B^T u^n(x) + C \\
&= \phi(0) + u^n(x)^T Q^\dagger u^n(x) + 2 \sum_{i=1}^{n} B_i \phi(\|x^i - x\|) + C \\
&\geq \phi(0) + u^n(x)^T Q^\dagger u^n(x) + 2 \sum_{B_i > 0} B_i \phi(d^{iR}) + 2 \sum_{B_i < 0} B_i \phi(D^{iR}) + C \\
&\geq \phi(0) + \Lambda'_a + 2 \sum_{B_i > 0} B_i \phi(d^{iR}) + 2 \sum_{B_i < 0} B_i \phi(D^{iR}) + C \\
&= \Lambda_a
\end{aligned}
$$

where the last but one inequality comes from $\phi(\|x^i - x\|) \in \left[ \phi(d^{iR}), \phi(D^{iR}) \right]$ and the last inequality comes from $Q^\dagger$ positive semidefinite and Theorem 3.33.

**2)** See proof for 2) in Corollary 3.32.

$\square$

**Corollary 3.35** *Let $\phi(r) = r^3$ be the* cubic spline *(monotonically increasing) and let $R = \left[ x^L, x^U \right] \subset \mathbb{R}_+^d$ an arbitrary box. Then for the* auxiliary problem *resp.* weighted auxiliary problem *in Definitions 3.3 resp. 3.6 the following lower bounds hold:*

**1)** *Consider the objective function*

$$a_n(x) := z^n(x)^T A^{-1} z^n(x) - \phi(0)$$

*of the* auxiliary problem. *We have:*

$$a_n(x) \geq \Lambda_a \quad for\ all \quad x \in R$$

*where*

$$\Lambda_a = \Lambda_a' + 2 \left[ \sum_{B_{i1}>0} B_{i1}\phi(d^{iR}) + \sum_{B_{i1}<0} B_{i1}\phi(D^{iR}) \right] + C_{11}$$

$$+ 2 \left[ \sum_{B_{ij}>0, j\geq 2} B_{ij}\phi(d^{iR})x_{j-1}^L + \sum_{B_{ij}<0, j\geq 2} B_{ij}\phi(D^{iR})x_{j-1}^U \right]$$

$$+ \left[ \sum_{C_{ij}>0, i,j\geq 2} C_{ij}x_{i-1}^L x_{j-1}^L + \sum_{C_{ij}<0, i,j\geq 2} C_{ij}x_{i-1}^U x_{j-1}^U \right]$$

$$+ 2 \left[ \sum_{C_{1j}>0, j\geq 2} C_{1j}x_{j-1}^L + \sum_{C_{1j}<0, j\geq 2} C_{1j}x_{j-1}^U \right] - \phi(0).$$

*and $\Lambda_a'$ comes from Theorem 3.33.*

**2)** *Consider the objective function*

$$w_n(x) := \frac{1}{[s_n(x) - f^*]^2} \left( z_n(x)^T A^{-1} z_n(x) - \phi(0) \right)$$

*of the* weighted auxiliary problem. *We have:*

$$w_n(x) \geq \Lambda_w \quad for\ all \quad x \in R$$

*where*

$$\Lambda_w = \frac{1}{[\Lambda_s - f^*]^2} \cdot \Lambda_a.$$

**Proof:**

**1)** Let $x \in R$. Recall that for the *cubic spline* we have $z^n(x) = (u^n(x)^T, \pi(x)^T)^T$ with $u^n(x) = (\phi(\|x^1 - x\|), \ldots, \phi(\|x^n - x\|))^T$ and $\pi(x) = (1, x_1, \ldots, x_d)^T$. For the auxiliary problem, we have:

$$
\begin{aligned}
a_n(x) &:= z^n(x)^T A^{-1} z^n(x) - \phi(0) \\
&= \begin{pmatrix} u^n(x)^T & \pi(x)^T \end{pmatrix} \begin{pmatrix} Q^\dagger & B \\ B^T & C \end{pmatrix} \begin{pmatrix} u^n(x) \\ \pi(x) \end{pmatrix} - \phi(0) \\
&= \begin{pmatrix} u^n(x)^T Q^\dagger + \pi(x)^T B^T & , & u^n(x)^T B + \pi(x)^T C \end{pmatrix} \begin{pmatrix} u^n(x) \\ \pi(x) \end{pmatrix} - \phi(0) \\
&= \left( u^n(x)^T Q^\dagger u^n(x) + \pi(x)^T B^T u^n(x) + u^n(x)^T B \pi(x) + \pi(x)^T C \pi(x) \right) - \phi(0) \\
&= u^n(x)^T Q^\dagger u^n(x) + 2 u^n(x)^T B \pi(x) + \pi(x)^T C \pi(x) - \phi(0) \\
&= u^n(x)^T Q^\dagger u^n(x) + 2 \sum_{i=1}^{n} \sum_{j=1}^{d+1} B_{ij} \phi(\|x^i - x\|) \pi_j(x) + \sum_{i=1}^{d+1} \sum_{j=1}^{d+1} C_{ij} \pi_i(x) \pi_j(x) - \phi(0) \\
&= u^n(x)^T Q^\dagger u^n(x) + 2 \sum_{i=1}^{n} B_{i1} \phi(\|x^i - x\|) + 2 \sum_{i=1}^{n} \sum_{j=2}^{d+1} B_{ij} \phi(\|x^i - x\|) x_{j-1} \\
&\quad + C_{11} \cdot 1 \cdot 1 + \sum_{i=2}^{d+1} \sum_{j=2}^{d+1} C_{ij} x_{i-1} x_{j-1} + 2 \sum_{j=2}^{d+1} C_{1j} x_{j-1} - \phi(0) \\
&\geq u^n(x)^T Q^\dagger u^n(x) + 2 \left[ \sum_{B_{i1}>0} B_{i1} \phi(d^{iR}) + \sum_{B_{i1}<0} B_{i1} \phi(D^{iR}) \right] + C_{11} \\
&\quad + 2 \sum_{i=1}^{n} \sum_{j=2}^{d+1} B_{ij} \phi(\|x^i - x\|) x_{j-1} + \sum_{i=2}^{d+1} \sum_{j=2}^{d+1} C_{ij} x_{i-1} x_{j-1} + 2 \sum_{j=2}^{d+1} C_{1j} x_{j-1} - \phi(0) \\
&\geq \Lambda_a' + 2 \left[ \sum_{B_{i1}>0} B_{i1} \phi(d^{iR}) + \sum_{B_{i1}<0} B_{i1} \phi(D^{iR}) \right] + C_{11} \\
&\quad + 2 \left[ \sum_{B_{ij}>0, j\geq 2} B_{ij} \phi(d^{iR}) x_{j-1}^L + \sum_{B_{ij}<0, j\geq 2} B_{ij} \phi(D^{iR}) x_{j-1}^U \right] \\
&\quad + \left[ \sum_{C_{ij}>0, i,j\geq 2} C_{ij} x_{i-1}^L x_{j-1}^L + \sum_{C_{ij}<0, i,j\geq 2} C_{ij} x_{i-1}^U x_{j-1}^U \right] \\
&\quad + 2 \left[ \sum_{C_{1j}>0, j\geq 2} C_{1j} x_{j-1}^L + \sum_{C_{1j}<0, j\geq 2} C_{1j} x_{j-1}^U \right] - \phi(0) \\
&= \Lambda_a
\end{aligned}
$$

where the last inequality comes from $\phi(\|x^i - x\|) \in \left[\phi(d^{iR}), \phi(D^{iR})\right]$ and Theorem 3.33.

**2)** See proof for 2) in Corollary 3.32.

$\square$

In the case of *thin plate splines* we can not use strict monotonicity like in the Corollaries 3.34 and 3.35. Nevertheless, with the notation $\phi_{min}^{iR}$ and $\phi_{max}^{iR}$ from Section 3.2.2.2, we can compute lower bounds in a similar way. Then the Corollary 3.36 is a direct consequence of Corollary 3.35.

**Corollary 3.36** *Let* $\phi(r) = r^2 \log r$ *be the thin plate spline and let* $R = \left[x^L, x^U\right] \subset \mathbb{R}_+^d$ *be an arbitrary box. Then for the auxiliary problem resp. weighted auxiliary problem in Definitions 3.3 resp. 3.6 the following lower bounds hold:*

**1)** *Consider the objective function*

$$a_n(x) := z^n(x)^T A^{-1} z^n(x) - \phi(0)$$

*of the auxiliary problem. We have:*

$$a_n(x) \geq \Lambda_a \quad \text{for all} \quad x \in R$$

*where*

$$\Lambda_a = \Lambda_a' + 2 \left[ \sum_{B_{i1}>0} B_{i1}\phi_{min}^{iR} + \sum_{B_{i1}<0} B_{i1}\phi_{max}^{iR} \right] + C_{11}$$

$$+ 2 \left[ \sum_{B_{ij}>0, j\geq 2} B_{ij}\phi_{min}^{iR} x_{j-1}^L + \sum_{B_{ij}<0, j\geq 2} B_{ij}\phi_{max}^{iR} x_{j-1}^U \right]$$

$$+ \left[ \sum_{C_{ij}>0, i,j\geq 2} C_{ij} x_{i-1}^L x_{j-1}^L + \sum_{C_{ij}<0, i,j\geq 2} C_{ij} x_{i-1}^U x_{j-1}^U \right]$$

$$+ 2 \left[ \sum_{C_{1j}>0, j\geq 2} C_{1j} x_{j-1}^L + \sum_{C_{1j}<0, j\geq 2} C_{1j} x_{j-1}^U \right] - \phi(0).$$

*and* $\lambda_a'$ *is the number from Theorem 3.33.*

**2)** *Consider the objective function*

$$w_n(x) := \frac{1}{[s_n(x) - f^*]^2} \left( z^n(x)^T A^{-1} z^n(x) - \phi(0) \right)$$

*of the* weighted auxiliary problem. *We have:*

$$w_n(x) \geq \Lambda_w \quad for\ all \quad x \in R$$

*where*

$$\Lambda_w = \frac{1}{[\Lambda_s - f^*]^2} \cdot \Lambda_a.$$

**Proof:**

The proof is analogous to that of Corollary 3.35. □


## 3.5  The Branch and Bound algorithm

In Sections 3.2 and 3.4 we developed the lower bounds $\Lambda_s$, $\Lambda_a$ and $\Lambda_w$ for the continuous functions $s_n$, $a_n$ and $w_n$ over a given rectangle $R = [x^L, x^U] \subset \mathbb{R}^d_+$. Here the lower bounds depend on several parameters. These parameters belong to two different types: The first type are parameters that do not depend on the rectangle. These are the the sample points $x^1, \ldots, x^n$, the matrix $A^{-1}$, the radial basis function $\phi$ and, where required, the value $f^*$. To keep our routine clearly represented, we will only note those parameters for input in our pseudocode, which are needed for the second class of parameters. The second class of parameters are those, which change with every rectangle. These are the distances $d^{iR}$ and $D^{iR}$, the numbers $\phi(d^{iR})$ and $\phi(D^{iR})$, which define a rectangle $V$ (see Remarks 3.11 and 3.12) and of course the rectangle $R$ itself, represented by the vectors $x^L$ and $x^U$. The next step will be to use these lower bounds within a Branch and Bound routine. For a Branch and Bound routine, the second class of parameters are of main interest, because these numbers have to be updated within the routine. To indicate that $\Lambda_g$ depends on the rectangle and a vector $u \in V$, we will use the notation $\Lambda_g(R, u)$, $u \in V$, instead of $\Lambda_g$ for any of the functions $g \in \{s_n, a_n, w_n\}$.

In addition to the lower bounds the main incredients within a Branch and Bound routine are the computation of upper bounds for the global optimum and the dividing of subsets, here: rectangles. The rough procedure is as follows: We start with the rectangle $R_0$ of the original problem and compute a lower bound and an upper bound for the function $g$ over $R_0$. If the lower bound is greater than the upper bound, then we are finished due to the fact that $R_0$ is empty. In case of nonemptiness we store the (global) upper bound and divide $R_0$ into two rectangles and, again, compute a lower and an upper bound for each of the two rectangles. We update the upper bound and compare it with both (local) lower bounds. We throw away rectangles, where the lower bound is greater than the upper bound and select a rectangle to be divided next. Here we use the "bisection rule", i.e. we choose the longest edge of the chosen rectangle and bisect this edge right in the middle and bisect the whole rectangle parallel to all other dimensions. After bisection we choose the next rectangle of our stack to be divided, and we iterate this process.

As the aim is to find and to bisect rectangles with a global minimum, we try to select the most promising rectangles to be divided next. Here we chose the rectangle with the smallest lower bound. To avoid that we store and search through a huge set of rectangles, a second aim is to quickly discard those rectangles, which obviously do not contain a global minimum. To do this it is beneficial to compute tight upper bounds. As we do not want to have much effort with the computation of upper bounds we will use a heuristic, see Algorithm 3.2, that will be explained in detail in Section 3.5.1.

Now it is possible to formulate the Branch and Bound algorithm.

### 3.5.1   The routine

Let $g : \mathbb{R}^n \to \mathbb{R}$ stand for one of the cheap objective functions $a_n$ of the *auxiliary problem*, or $w_n$ of the *weighted auxiliary problem*, or the interpolating function $s_n$ itself over the box $R_0 = \left[x^L, x^U\right] \subset \mathbb{R}^d_+$. Let $x^1, \ldots, x^n$ be the already sampled points and let $\Lambda_g(R, u)$ (resp. $\Lambda^1_g(R, u)$ or $\Lambda^2_g(R, u)$), $u \in V$, be the lower bound for $g$, which are given by Corollaries 3.32, 3.34 and 3.35. Then we can use Algorithm 3.1 for minimizing the given cheap function. Algorithm 3.1 is based on a routine taken from [20].

The upper bounds for our cheap function $g$ which are used in Algorithm 3.1, are computed with a heuristic called "Improving Hit and Run (IHR)" [47], which we implemented as in Algorithm 3.2. The procedure roughly works as follows: we start with a given point $v \in R_0$ and move into a random direction $d$. Here $d$ is uniformly distributed over the set $Z_{R_0}(v)$, which contains all feasible directions in $v$. Here "feasible direction" means that it is possible to move a positive stepsize in this direction without leaving the rectangle $R_0$. Of course, if $v$ is an inner point of $R_0$, then all directions are feasible. To identify the directions which are infeasible, we have to check if $v$ lies on the boundary of $R_0$. After chosing the direction $d$ we chose a point on the line set between $v$ and the boundary of $R_0$, by using a (positive) random

---
**Algorithm 3.1:** Branch and Bound
---

**Input:** Vectors $x^L$ and $x^U$, fixed evaluation points $x^1, \ldots, x^n$, function $\phi$
cheap objective function $g : \mathbb{R}^d \to \mathbb{R}$,
lower bound function $\Lambda_g(R, u)_{u \in V} : \mathbb{R}_+^d \times \mathbb{R}_+^n \to \mathbb{R}$ (see Rem. 3.11 or 3.12 for $V$).

**Output:** Optimum $v$ and optimal objective function value $g(v)$.

**Initialization:**

Set $a \leftarrow x^L$, $b \leftarrow x^U$, and $Rec := R_0 = [a, b]$.

Compute distances $d^{iR_0}$ and $D^{iR_0}$. Define $V_0$ as in Remark 3.11 resp. 3.12.

Set $ub \leftarrow \min\{g(a), g(b)\}$, $v \leftarrow \operatorname{argmin}\{g(a), g(b)\}$, $N \leftarrow$ Number of iterations for IHR, and execute IHR$(v, ub, R_0, g, N)$.

Set $v \leftarrow$ output1(IHR), $ub \leftarrow$ output2(IHR), $\mathcal{B} \leftarrow \{Rec\}$ and
$lb \leftarrow lb(Rec) = \Lambda_g(Rec, u)(= \Lambda_g(R_0, u))$ with $u \in V$, $stop \leftarrow false$.

**while** $stop = false$ **do**

    **if** $ub = lb$ **then**

        |   $stop \leftarrow true$ ($v$ is optimal solution and $ub$ the opt. obj. function value)

    **else**

        Compute $b_j - a_j = \max\{b_i - a_i : i = 1, \ldots, d\}$ and set
            $a^1 \leftarrow a, b^1 \leftarrow (b_1, \ldots, b_{j-1}, (b_j + a_j)/2, b_{j+1}, \ldots, b_d)^T$
            $a^2 \leftarrow (a_1, \ldots, a_{j-1}, (b_j + a_j)/2, a_{j+1}, \ldots, a_d)^T, b^2 \leftarrow b$
            $R_1 \leftarrow \{x : a^1 \le x \le b^1\}, R_2 \leftarrow \{x : a^2 \le x \le b^2\}$.
        Compute distances $d^{iR_1}$ and $D^{iR_1}$ for $i = 1, \ldots, n$.
        Compute distances $d^{iR_2}$ and $D^{iR_2}$ for $i = 1, \ldots, n$.
        Define $V_1$ and $V_2$ as in Remark 3.11 resp. Remark 3.12.
        Set
            $lb(R_1) \leftarrow \max\{lb(R), \Lambda_g(R_1, u) \text{ with } u \in V_1\}$;
            $lb(R_2) \leftarrow \max\{lb(R), \Lambda_g(R_2, u) \text{ with } u \in V_2\}$;
            $ub \leftarrow \min\{ub, g(b^1), g(a^2)\} = \min\{g(v), g(b^1), g(a^2)\}$;
            $v \leftarrow \arg(ub)$.
        Execute IHR$(v, ub, R_0, g, N)$ and update
            $v \leftarrow$ IHR-Output(component1);
            $ub \leftarrow$ IHR-Output(component2);
            $\mathcal{B} \leftarrow \{R \in (\mathcal{B} \setminus \{Rec\} \cup \{R_1, R_2\}) \text{ with } lb(R) < ub\}$;
        
$$lb \leftarrow \begin{cases} \min\{lb(R) : R \in \mathcal{B}\} & \text{if } \mathcal{B} \ne \emptyset \\ ub & \text{if } \mathcal{B} = \emptyset. \end{cases}$$
        Choose $Rec \in \mathcal{B}$ satisfying $lb(Rec) = lb$.
        Update $a$ and $b$ such that $Rec = [a, b]$.

    **end**

**end**

step size $\lambda$ in direction $d$. We evaluate the function $g$ at this point and accept the new candidate point only if it is improving in the objective function value. In detail this means: if the function value is smaller than $g(v)$, we use the new point for chosing again a random direction to go. Otherwise we use $v$ for chosing a random direction to go. We repeat this procedure $N$ times.

---

**Algorithm 3.2:** Improving Hit and Run (IHR)

---

**Input:** $v, \mathrm{ub}, R_0, g, N$

**Output:** output1 = improved $v$ and

output2 = improved ub.

**Initialization:**

Set

$D_0 \leftarrow Z_{R_0}(v)$ (Set of feasible directions of $v \in R_0$.);

$x^0 \leftarrow v, \quad y^0 \leftarrow \mathrm{ub}, \quad k \leftarrow 0$.

**while** $k < N$ **do**

Generate a random direction $d^k$ uniformly distributed over $D_k$.

Sample a candidate point $w^{k+1}$ uniformly over the line set

$$L_k = \left\{ x \in R_0 \mid x = x^k + \lambda d^k \text{ with } \lambda \text{ a real scalar} \right\}.$$

Set

$$x^{k+1} \leftarrow \begin{cases} w^{k+1} & \text{if } g(w^{k+1}) < \mathrm{ub}, \\ v & \text{otherwise;} \end{cases}$$

$y^{k+1} \leftarrow g(x^{k+1})$;

$D_{k+1} \leftarrow Z_{R_0}(x^{k+1})$;

$k \leftarrow k + 1$.

**end**

Set

output1 $\leftarrow x^{k+1}$;

output2 $\leftarrow y^{k+1}$.

---

## 3.5.2 Convergence of the Branch and Bound algorithm

Before we prove the convergence of Algorithm 3.1, we introduce the important property of *exhaustiveness* that is needed for the subdivision of rectangles. For this we note that a sequence of rectangles is called "nested" if every rectangle of this sequence is a subset of its predecessor, i.e. $R_k \supseteq R_{k+1}$ for all $k$. Exhaustiveness means that the diameters of every infinite nested sequence of rectangles converges to zero. With the definition of the *diameter of a rectangle* $R = [x^L, x^U]$,

$$\delta(R) := \max \{\|x - y\| : x, y \in R\} = \|x^L - x^U\|$$

we can describe exhaustiveness more formally as follows: Exhaustiveness means that $\delta(R_k) \to 0$ for $k \to \infty$ for every nested sequence $\{R_k\}$. It is intuitively clear that the bisection rule we use in Algorithm 3.1 leads to exhaustive sequences of rectangles. For the formal proof we refer to Chapter 4 resp. [20].

Next, we show that Algorithm 3.1 converges for $k \to \infty$. To distinguish the rectangles, bounds, points and parameters on the $k$-th stage of the algorithm, we will refer to $R_k$ instead of $R$ etc.

**Theorem 3.37** *If Algorithm 3.1 is infinite, then*

$$\mathrm{lb} = \lim_{k \to \infty} \mathrm{lb}_k = \lim_{k \to \infty} g(v^k) = \lim_{k \to \infty} \mathrm{ub}_k = \mathrm{ub},$$

*and every accumulation point $v^*$ of the sequence $\{v^k\}$ is an optimal solution of the problem* $\min \{g(x) : x \in [x^L, x^U]\}$.

**Proof:** It suffices to show (see [20, Theorem 3.8]) that for every decreasing sequence $R_{k_q}$, $R_{k_q} \supset R_{k_{q+1}}$, of successively refined rectangles the corresponding bounds satisfy

$$\lim_{q \to \infty} (\mathrm{ub}_{k_q} - \mathrm{lb}_{k_q}) = 0.$$

The idea is to construct a sequence of upper bounds, $\mathrm{ub}(R_{k_q})$, which overestimates $\mathrm{ub}_{k_q}$, and a sequence of lower bounds, $\mathrm{lb}'(R_{k_q})$, which underestimates $\mathrm{lb}_{k_q}$ for every $q$. If both sequences have the same limit point for $q \to \infty$, the proof is done.

Let $A_{k_q} = \{v^{k_q}, b^{k_q(1)}, a^{k_q(2)}\} \cap R_{k_q}$ be the set of evaluated points in $R_{k_q}$. The set $A_{k_q}$ is of course nonempty because the points $b^{k_q(1)}$ and $a^{k_q(2)}$ result from the bisection on stage $k_q$ of the algorithm and lie in $R_{k_q}$.

Let

$$\mathrm{ub}(R_{k_q}) := \min \{g(x) : x \in A_{k_q}\}$$

be an upper bound for the rectangle $R_{k_q}$. Since $A_{k_q} \subseteq \left\{ v^{k_q}, b^{k_q(1)}, a^{k_q(2)} \right\}$ and $\mathrm{ub}_{k_q} = g(v^{k_q}) \leq \left\{ g(x) : x \in \left\{ v^{k_q}, b^{k_q(1)}, a^{k_q(2)} \right\} \right\}$ we have that the sequence $\mathrm{ub}(R_{k_q})$ overestimates the sequence $\mathrm{ub}_{k_q}$:

$$\mathrm{ub}(R_{k_q}) \geq \mathrm{ub}_{k_q} \quad \text{for all } q.$$

Due to the fact that the bisection method in our algorithm is exhaustive (see [20, Lemma 5.1]), we conclude that

$$\lim_{q \to \infty} R_{k_q} = \bigcap_{q=1}^{\infty} R_{k_q} = \{s\} \tag{3.10}$$

for some point $s$ and therefore

$$\lim_{q \to \infty} \mathrm{ub}(R_{k_q}) = g(s).$$

A lower bound for the rectangle $R_{k_q}$ is

$$\mathrm{lb}'(R_{k_q}) := \Lambda_g(R_{k_q}, u^{k_q}),$$

where $u^{k_q} = \mathrm{argmin} \left\{ u^T Q^\dagger u \mid u \in V_{k_q} \right\}$.

What we want to do next is to show that the relation

$$\mathrm{lb}'(R_{k_q}) \leq \mathrm{lb}_{k_q} \quad \text{for all } q$$

holds. Therefore we have to understand that the sequence $R_{k_q}$ is a subsequence of the sequence of successfully bisected rectangles $R_k$. Due to this fact the rectangle chosen to bisect next in Algorithm 3.1 is the one with the smallest lower bound among the currently stored rectangles. Therefore the relation above holds.

Due to (3.10) the mininum and maximum distances between an arbitrary point $x^i$ and $R_{k_q}$ converge to the distance between $x^i$ and $s$, i.e.

$$\lim_{q \to \infty} d^{ik_q} = \lim_{q \to \infty} D^{ik_q} = \|x^i - s\|.$$

In addition, also as a consequence of (3.10), the set $V_{k_q}$ shrinks to the point $(\phi(\|x^1 - s\|), \dots, \phi(\|x^n - s\|))$ and therefore we have $u^{k_q} \to u = (\phi(\|x^1 - s\|), \dots, \phi(\|x^n - s\|))$, i.e.

$$\lim_{q \to \infty} (u^{k_q})^T Q^\dagger u^{k_q} = (\phi(\|x^1 - s\|), \dots, \phi(\|x^n - s\|))^T Q^\dagger (\phi(\|x^1 - s\|), \dots, \phi(\|x^n - s\|))$$

$$= u^T Q^\dagger u.$$

Furthermore, we have that (with $s = (s_1, \ldots, s_d)$ and $R_{k_q} = [x^{L_{k_q}}, x^{U_{k_q}}]$)

$$\lim_{q \to \infty} x_j^{L_{k_q}} = s_j \text{ for } j = 1, \ldots, d.$$

Now we can replace the minimum and maximum distances in $\Lambda_g(R_{k_q}, u^{k_q})$ as well as the polynomial and mixed polynomial parts and get

$$\lim_{q \to \infty} \mathrm{lb}'(R_{k_q}) = g(s).$$

Using the relations

$$\mathrm{lb}'(R_{k_q}) \le \mathrm{lb}_{k_q} \le \mathrm{ub}_{k_q} \le \mathrm{ub}(R_{k_q}) \quad \text{for all } q,$$

and letting $q \to \infty$, the allegation follows. $\qquad \square$

# Chapter 4

# Box partitioning

## 4.1 The problem

One of the central steps within every iteration of the Branch and Bound routine described in Chapter 3 is the bisection of the chosen rectangle into two subrectangles. In Algorithm 3.1 we divided the rectangles right in the middle of the longest edge and it was intuitively clear, that this bisection rule leads to exhaustivness. The property *exhaustiveness* was explained in the beginning of Section 3.5.2. We recall the formal definition: Exhaustiveness means that $\delta(R_k) \to 0$ for $k \to \infty$ for every nested sequence $\{R_k\}$, where $\delta(R)$ is the diameter of the rectangle $R$. Of course, there are many different possible bisection strategies, but not each of them leads to the required division rule that constructs an exhaustive sequence of rectangles and ensures the convergence of the whole algorithm. Proving exhaustiveness requires information on the relation between the diameter $\delta(R_{k+1})$ and the diameter $\delta(R_k)$. The question now is what other division rules exist for bisection, which guarantee that the sequence of rectangles is exhaustive.

Bisection of simplices was investigated by several authors [19, 8]. Horst [19], for example, presents a formula that puts the diameter of a simplex in relation to the diameter of the predecessor and, finally, gives partitioning rules that ensure the contraction of the sequence of diameters to zero, which means exhaustiveness. In the following sections, we will develop a partitioning rule for the bisection of rectangles similar to the one Horst presented in his paper [19] for simplices.

One important piece of information for the development of such a formula is the position of the vertices relative to the midpoint of the rectangle. This obvious fact is stated in Lemma 4.1.

**Lemma 4.1** *Let* $R = [x^L, x^U] \subset \mathbb{R}^d$ *be a rectangle. We define the midpoint* $m$ *of* $R$ *as*

$$m := \tfrac{1}{2}(x^L + x^U).$$

*Then the distance between any vertex* $x^i$ *of* $R$ *and* $m$ *equals* $\frac{1}{2}\delta(R)$.

With Lemma 4.1 we can conclude that $B_{\frac{1}{2}\delta(R)}(m)$ is the smallest ball centered in $m$ containing $R$. In Figure 4.1 we see the smallest ball for a 2-dimensional rectangle.

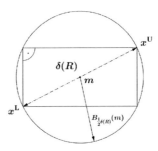

Figure 4.1: Smallest ball containing $R$ in the 2-dimensional case.

## 4.2 General bisection of $d$-dimensional rectangles

Now assume that the rectangle $R_{k+1}$ is constructed from $R_k$ by some construction rule. The following section focusses on the computation of (the possible maximum value of) $\delta(R_{k+1})$ from the given rectangle $R_k$ with diameter $\delta(R_k)$. To keep the calculation easy, we follow some rules for division, without loss of generality: For the computation of $\delta(R_{k+1})$ we will always consider

- the longest edge of $R_k$ to be bisected and, after bisecting,

- we consider the diameter of the larger rectangle.

Theoretically this causes no problem as the aim is to overestimate the diameter of the new rectangle in order to prove the exhaustiveness. Therefore all other sequences of (smaller) diameters will of course converge, too. From practical point of view we note that

- we can always, without loss of generality, use the longest edge that is adjacent with $x^U$ for bisection and

- choosing a bisection point between $m$ and $x^U$ ensures that the bigger of the two subrectangles has $x^L$ as its lower left vertex.

We will develop a geometrical construction, which ensures that we comply with these rules of division, in the Sections 4.2.1 and 4.2.2.

## 4.2.1 The regular $d$-dimensional rectangle

Consider the unit cube $[0,1]^d$ with $x^L = \mathbb{0}_d$ and $x^U = \mathbb{1}_d$. Let $\bar{x}^j$ be any vertex of $[0,1]^d$ adjacent to $\mathbb{1}_d$. Because the cube is a cartesian rectangle, we have that $\bar{x}^j$ is an element of the set

$$
\left\{
\begin{pmatrix} 0 \\ 1 \\ 1 \\ \vdots \\ 1 \end{pmatrix},
\begin{pmatrix} 1 \\ 0 \\ 1 \\ \vdots \\ 1 \end{pmatrix},
\cdots,
\begin{pmatrix} 1 \\ 1 \\ \vdots \\ 1 \\ 0 \end{pmatrix}
\right\}.
$$

Then $\|x^U - \bar{x}^j\| = 1$ and $\|x^L - x^U\| = \sqrt{d}$ and $\|x^L - \bar{x}^j\| = \sqrt{d-1}$. Figure 4.2 shows the triangle between $x^L$, $\bar{x}^j$ and $x^U$.

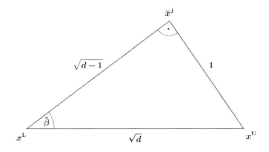

Figure 4.2: The triangle $x^L, x^U, \bar{x}^j$.

By the theorem of Pythagoras, we get that the angle between $x^L - \bar{x}^j$ and $x^U - \bar{x}^j$ is 90°. Let $\bar{\beta}$ be the angle between the vectors $\bar{x}^j - x^L$ and $x^U - x^L$. By standard trigonometric results, we have

$$
\sin \bar{\beta} = \frac{1}{\sqrt{d}}. \tag{4.1}
$$

Note that in Figure 4.2 the edge between $\bar{x}^j$ and $x^U$ not only measures the distance between $\bar{x}^j$ and $x^U$, but corresponds to a real existent (longest) edge of the cube $[0,1]^d$, too. In contrast, the edge between $\bar{x}^j$ and $x^L$ does not correspond to an edge in the considered cube $[0,1]^d$, except for the 2- and 1-dimensional cases.

In order to compare the cube with general rectangles we will represent the right angularity in Figure 4.2 additionally by using the Theorem of Thales, which says that in a rectangular triangle the vertex corresponding to the 90°-angle lies on a semicircle, where the diameter is given by the hypothenuse. I.e. we can inscribe the triangle into a semicircle with diameter $\delta(\text{cube}) = \sqrt{d}$ and midpoint $m$ (see Figure 4.3).

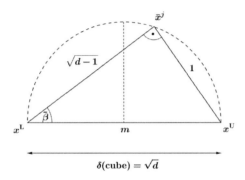

Figure 4.3: Thales Theorem: The triangle $x^L, x^U, \bar{x}^j$ inscribed in a semicircle.

Next let us consider a general rectangle $R$ with $\delta(R) = \sqrt{d}$ and the $d$ paraxial edges $a_1, \ldots, a_d$, which are adjacent to $x^U$. Moreover, due to the perpendicularity of the rectangle, we have by using Pythagoras multiple times

$$\sum_{i=1}^{d} \|a_i\|^2 = d. \tag{4.2}$$

In the special case of the unit cube we have the unit length $\|a_1\| = \ldots = \|a_d\| = 1$ and the trivial equation $d \cdot 1^2 = d$. Therefore, with (4.2), we conclude that for every rectangle $R \neq [0,1]^d$ with $\delta(R) = \sqrt{d}$ the longest edge, say $a_j$, is longer than the longest edge of $[0,1]^d$. With $a := \|a_j\|$ this means

$$a > 1.$$

Now let us rewrite equation (4.2) as the sum

$$d = \underbrace{\sum_{i \neq j} \|a_i\|^2}_{=:c^2} + \underbrace{\|a_j\|^2}_{=a^2} = c^2 + a^2$$

and extend the Figure 4.3 by inscribing the triangle generated by $x^L$, $x^j$, which is a vertex of $R$ that is adjacent to $x^U$, and $x^U$. This triangle now represents the rectangle $R$ (see Figure 4.4).

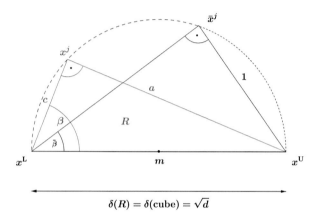

$$\delta(R) = \delta(\text{cube}) = \sqrt{d}$$

Figure 4.4: Comparison of the longest edges of a rectangle $R$ and the cube $[0,1]^d$, both with $\delta(R) = \delta(cube) = \sqrt{d}$.

Figure 4.4 now illustrates the fact that for an arbitrary rectangle $R$ with $\delta(R) = \sqrt{d}$ the vertex $x^j$, which is adjacent to $x^U$ via a longest edge $a_j$, always lies on that part of the semicircle which is on the left side of $\bar{x}^j$ (red dotted line). This fact now leads to the consequence that $\beta \geq \bar{\beta}$. An upper bound for $\beta$ is given by 90° if we move $x^j$ as far as possible to the left on the semicircle. Hence we have, by using (4.1) for the first inequality,

$$\sqrt{\tfrac{1}{d}} \leq \sin \beta \leq 1 \tag{4.3}$$

for an arbitrary rectangle $R$ with $\delta(R) = \sqrt{d}$.

Since multiplying all lengths in the Figures 4.2 - 4.4 by a constant factor does not change the equation (4.1), the estimation (4.3) holds for rectangles and cubes with arbitrary diameters.

In Section 4.2.2 we will deduce a second consequence of $x^j$ lying on the left hand circle of $\bar{x}^j$ in Figure 4.4. This second consequence contains a rough geometrical estimation of the diameter of a rectangle constructed from $R$ by bisecting a longest edge.

## 4.2.2  A geometrical estimation

With Section 4.2.1 we are able to deduce an upper bound for the diameter of a successor rectangle after bisection. For this let $w$ be a bisection point lying on the diagonal between $x^L$ and $x^U$, more precisely between $m$ and $x^U$. In order to bisect the longest edge of the rectangle $R$ respectively the cube, we draw the two lines that are perpendicular to the longest edge of $R$ respectively to the longest edge of the cube and receive the bisection points $w^R$ and $w^{\text{cube}}$ (see Figure 4.5).

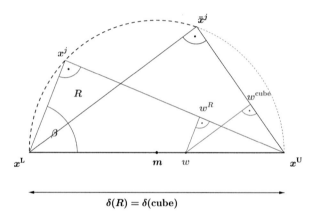

Figure 4.5: Bisection with $w$.

After bisection we receive two subrectangles of $R$ and the two subrectangles of the cube. Let us call the larger one $R_{suc}$ respectively $\text{cube}_{suc}$. The next step is to compare the diameters $\delta(R_{suc}) = \|x^L - w^R\|$ and $\delta(\text{cube}_{suc}) = \|x^L - w^{cube}\|$, see Figure 4.6.

Then it is clear that

$$\delta(R_{suc}) \leq \delta(\text{cube}_{suc}).$$

In summary we can make the following observation: If we consider all possible shapes of rectangles with the same diameter, the worst case for the division of a rectangle is the one where all of its edges have the same length, i.e. if we consider the unit cube.

70

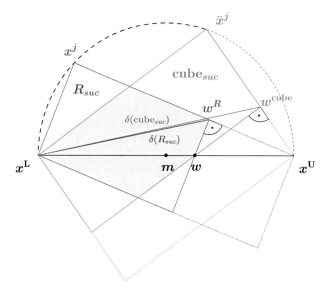

Figure 4.6: Successive rectangles.

Here the "worst case" means: If we bisect all rectangles with the same diameter at the same bisection point, the successor rectangle of the unit cube is the largest one, compared to the successor rectangles of all other bisected rectangles.

Therefore we have a first geometrical estimation for the diameter of the successor rectangle after bisection. In addition, we have a justification why we consider the (unit) cube as the extreme case of a rectangle. With these notations the "division rules" formulated at the beginning of Section 4.2 can be incorporated in Section 4.2.3.

In addition, we finally deduce the crucial theorem that contains the exact bisection rule for ensuring the exhaustiveness of a sequence of bisected rectangles in Section 4.2.3

## 4.2.3   General Rectangle

As a result of the previous sections we can formulate Lemma 4.2.

**Lemma 4.2** *Let $R = [x^L, x^U] \subset \mathbb{R}^d$ be a rectangle with $d \geq 1$ and let $\delta(R)$ denote the diameter of $R$. We construct the new rectangle $R_{suc}$ by bisecting a longest edge*

71

of R. For this, let $x^j$ be a vertex of R that is adjacent to $x^U$ via this longest edge, say $a_j$. Let w be a point on the diagonal constructed as

$$w = \lambda x^L + (1 - \lambda)x^U \ \text{with} \ 0 \le \lambda \le \frac{1}{2}.$$

Furthermore,

$$w^R := \lambda x^j + (1 - \lambda)x^U \ \text{with} \ 0 \le \lambda \le \frac{1}{2}$$

is, by standard trigonometric results, the appropriate intersection point on the edge $a_j$ (see also Figure 4.6).

Then we can conlude that

$$\delta(R_{suc}) \le \sqrt{1 + \tfrac{1}{d}\lambda^2 - \tfrac{2}{d}\lambda}\,\delta(R).$$

Before we prove Lemma 4.2, which mainly consists of the construction of the diameter $\delta(R_{suc})$, we take a look at the geometrical representation of the considered rectangles: As before, let $R = [x^L, x^U]$ be the rectangle that we are going to bisect. To construct the diameter $\delta(R_{suc})$ of the next rectangle $R_{suc}$, it suffices to consider the semicircle through $x^L$, $x^U$, and $x^j$ (see Sections 4.2.1 and 4.2.2 and Figure 4.7):

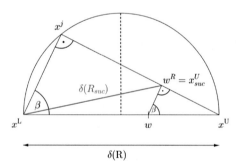

Figure 4.7: The diameter of $R_{suc}$.

By construction, $R_{suc} = [x^L, w^R]$, therefore we set $x^U_{suc} = w^R$. In Figure 4.7 $x^j$ is the vertex, that is adjacent to $x^U$ via a longest edge. Now we take a detailed look at the triangle with the vertices $w$, $x^U_{suc}$ and $x^U$ in Figure 4.7. For this let us consider Figure 4.8, which shows the triangle in Figure 4.7 in more detail.

As we will see shortly both the values $q$ and $h$ in Figure 4.8 are required for the proof of Lemma 4.2 and for the computation of $\delta(R_{suc})$.

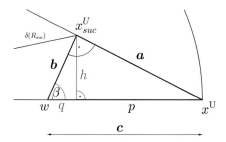

Figure 4.8: The triangle $w$, $x_{suc}^U$, $x^U$.

**Lemma 4.3** *For the triangle in Figure 4.8, we have*

$$b = c \cdot \cos \beta,$$
$$q = c \cdot \cos^2 \beta,$$
$$h = c \cdot \cos \beta \sin \beta.$$

**Proof of Lemma 4.3:**

By standard trigonometric results, we have

$$\frac{b}{c} = \cos \beta \quad \Leftrightarrow \quad b = c \cdot \cos \beta,$$

and

$$b^2 = q \cdot c \quad \Leftrightarrow \quad q = \frac{b^2}{c} = \frac{(c \cdot \cos \beta)^2}{c} = c \cdot \cos^2 \beta. \tag{4.4}$$

Finally, the right triangle altitude theorem gives

$$
\begin{aligned}
h^2 &= q \cdot p \\
&= c \cos^2 \beta \cdot (c - c \cos^2 \beta) \\
&= c^2 \cos^2 \beta (1 - \cos^2 \beta) \\
&= c^2 \cos^2 \beta \sin^2 \beta
\end{aligned}
$$

which is equivalent to

$$h = c \cdot \cos \beta \sin \beta. \tag{4.5}$$

$\square$

With Lemma 4.3 we are now able to prove Lemma 4.2.

**Proof of Lemma 4.2:**

Let us consider Figure 4.7. Note that from $w = \lambda x^L + (1 - \lambda)x^U$, we get that

$$\|w - x^L\| = (1 - \lambda)\|x^U - x^L\| = (1 - \lambda)\delta(R)$$

and

$$\|x^U - w\| = \lambda \|x^U - x^L\| = \lambda \delta(R).$$

Due to the theorem of Thales and with both (4.4) and (4.5) we are now able to compute $\delta(R_{suc})$ as follows:

$$
\begin{aligned}
\delta^2(R_{suc}) &= \|x_{suc}^U - x^L\|^2 \\
&= (\|w - x^L\| + q)^2 + h^2 \\
&= (\|w - x^L\| + \cos^2 \beta \|x^U - w\|)^2 + (\cos \beta \sin \beta \cdot \|x^U - w\|)^2 \\
&= \left[(1 - \lambda)\delta(R) + \lambda \cos^2 \beta \delta(R)\right]^2 + \lambda^2 \cos^2 \beta \sin^2 \beta \delta^2(R) \\
&= (1 - \lambda)^2 \delta^2(R) + 2(1 - \lambda)\lambda \cos^2 \beta \delta(R)^2 + \lambda^2 \cos^4 \beta \delta^2(R) \\
&\quad + \lambda^2 \cos^2 \beta \sin^2 \beta \delta^2(R) \\
&= \delta^2(R) \left[(1 - \lambda)^2 + 2(1 - \lambda)\lambda \cos^2 \beta + \lambda^2 \cos^4 \beta + \lambda^2 \cos^2 \beta \sin^2 \beta\right] \\
&= \delta^2(R) \left[1 - 2\lambda + \lambda^2 + 2\lambda \cos^2 \beta - 2\lambda^2 \cos^2 \beta + \lambda^2 \cos^4 \beta + \lambda^2 \cos^2 \beta \sin^2 \beta\right] \\
&= \delta^2(R)[1 + \lambda^2 + \lambda^2 \cos^2 \beta \underbrace{(-2 + \cos^2 \beta + \sin^2 \beta)}_{=-1} + 2\lambda \underbrace{(\cos^2 \beta - 1)}_{=-\sin^2 \beta}] \\
&= \delta^2(R)[1 + \lambda^2 \underbrace{(1 - \cos^2 \beta)}_{=\sin^2 \beta} - 2\lambda \sin^2 \beta] \\
&= \delta^2(R) \left[1 + \sin^2 \beta (\lambda^2 - 2\lambda)\right]
\end{aligned}
$$

With the condition (4.3) for $\beta$, that is $\sin^2 \beta \geq \frac{1}{d}$, and due to $\lambda^2 - 2\lambda \leq 0$ finally the estimation is

$$\delta^2(R_{suc}) \leq \delta^2(R) \left[1 + \frac{1}{d}\lambda^2 - \frac{2}{d}\lambda\right].$$

$\square$

**Theorem 4.4** *Let $\{R_k\}_{k \in \mathbb{N}}$ be a nested sequence of rectangles in $\mathbb{R}^d$ with $d \geq 1$, where we construct $R_{k+1}$ by using the bisection rule proposed in Lemma 4.2 with the bisection point*

$$w_k = \lambda_k x_k^L + (1 - \lambda_k)x_k^U \text{ with } 0 < c \leq \lambda_k \leq \frac{1}{2}$$

*for some $0 < c < \frac{1}{2}$.*

*Then we can conlude that*

$$\lim_{k \to \infty} \delta(R_k) = 0.$$

**Proof:**

Since $d \geq 1$, $c \in (0, \frac{1}{2})$, and $0 < c \leq \lambda_k \leq \frac{1}{2}$, we get

$$\frac{1}{2} \leq \sqrt{\frac{4d - 3}{4d}} \leq \sqrt{1 + \frac{1}{d}\lambda_k^2 - \frac{2}{d}\lambda_k} \leq \sqrt{1 + \frac{1}{d}c^2 - \frac{2}{d}c} < 1, \tag{4.6}$$

where the first inequality comes from $d \geq 1$, the second inequality is due to $\lambda_k \leq \frac{1}{2}$, the third inequality follows from $\lambda_k \geq c$, and the last inequality comes from $0 < c \leq \frac{1}{2}$.

Due to Lemma 4.2 and (4.6) we have that

$$\frac{\delta(R_{k+1})}{\delta(R_k)} \leq \sqrt{1 + \frac{1}{d}\lambda_k^2 - \frac{2}{d}\lambda_k} \leq \sqrt{1 + \frac{1}{d}c^2 - \frac{2}{d}c}. \tag{4.7}$$

Since the right hand side in (4.7) is a fixed number strictly between 0 and 1, and since all $\delta(R_k)$ are positive, it follows that $\delta(R_k) \to 0$ as $k \to \infty$, which proves the assertion.

$\square$

# Chapter 5

# General convex constraints

In the previous chapters we considered the problem

$$\min f(x) \quad s.t. \, x \in R \subset \mathbb{R}^d,$$

where, without loss of generality, $R$ was assumed to be the unit box $[0,1]^d$ and $f$ was an expensive objective function. We incorporated a RBF algorithm containing special subproblems. The structure of these subproblems was similar to the original problem, except for the objective, which was replaced by a "cheap" function $g$. This enabled us to develop an appropriate Branch and Bound algorithm to solve these subproblems.

Now we are going to expand the global optimization problem by general convex constraints. We describe these convex constraints with convex functions $h_j(x)$, $j = 1, \ldots, m$. That is, we consider the problem

$$\min f(x)$$
$$s.t. \quad x \in [0,1]^d \tag{5.1}$$
$$h_j(x) \leq 0 \text{ for } j = 1, \ldots, m,$$

where $h_j : \mathbb{R}^d \to \mathbb{R} \, (j = 1, \ldots, m)$ are convex functions.

Within the Branch and Bound algorithm one important task was to find lower and upper bounds for a function $g$ over a subrectangle $R$ of $[0,1]^d$. Now we have to expand the box conditions by the convex constraints. That is, we now consider problems of the form

$$\min g(x)$$
$$s.t. \quad x \in R \tag{5.2}$$
$$h_j(x) \leq 0 \text{ for } j = 1, \ldots, m,$$

for a given subrectangle $R \subset [0,1]^d$.

The Branch and Bound Algorithm 3.1 we proposed in the previous chapter only works with box constraints. That is, the starting rectangle and every rectangle we get by subdivision are automatically feasible. Therefore, there was no need to test whether or not a rectangle is feasible. With the additional convex constraints

$$h_i(x) \leq 0 \quad \text{for } i = 1, \ldots, m$$

it is not even certain whether the starting set

$$\left\{ x \in [0,1]^d \,|\, h_i(x) \leq 0 \text{ for } i = 1, \ldots, m \right\}$$

is empty or not. For our tests we will assume that the starting set is nonempty. Therefore we have to decide for every $R$, except for the starting rectangle, whether the set

$$R \cap \left\{ x \in \mathbb{R}^d : h_i(x) \leq 0 \text{ for } i = 1, \ldots, m \right\}$$

is empty or not. Due to the convexity of the constraint set, this can easily be done by solving a convex minimization problem. In our case we will solve the problem by minimizing a trivial function over the given convex set, i.e. we solve the convex problem

$$\min x_1$$
$$s.t. \quad x \in R$$
$$\qquad h_j(x) \leq 0 \text{ for } j = 1, \ldots, m,$$

for the start set and at every iteration, where we get two new rectangles after dividing. We call this test for the validity of $R$ "ValRect" and its pseudo code is presented by Algorithm 5.1. If the return value of this procedure is "invalid", then the constraint set is empty and the corresponding rectangle will be deleted from the set of further considered rectangles. Note, that because of testing the starting rectangle, at least one of the two rectangles after subdivision will remain in the list of considered rectangles.

A second problem concerns the desire to keep the search area as small as possible, that is, to cut off areas which certainly do not contain valid points. This problem will also be tackled by solving convex optimization problems, see Section 5.1.

## 5.1 Scaling down rectangles

In the original Branch and Bound algorithm we always worked with rectangles that were completely contained in the feasible set. Therefore there was no need to shrink

**Algorithm 5.1:** ValRect

**Input**: $R, h_1(x), \ldots, h_m(x)$

**Output**: valid/invalid

Solve

$\quad \min x_1$ s.t. $x \in R \cap \left\{ x \in \mathbb{R}^d : h_i(x) \leq 0 \text{ for } i = 1, \ldots, m \right\}$

**if** *solution exists* **then**

$\quad$ **Output** ←valid

**else**

$\quad$ **Output** ←invalid

**end if**

the rectangles we considered. For a Branch and Bound algorithm for problem (5.1) the considered rectangles now potentially have parts that could be cut off. We will incorporate this by scaling the rectangles on every single dimension. This scaling keeps the search area small and, as a consequence, speeds up the routine of approximating lower and upper bounds. This scaling works as follows: Given a rectangle $R = [a, b]$ and the convex functions $h_1(x), \ldots, h_m(x)$ we will, for $k = 1, \ldots, d$, solve the problems

$$\tilde{a}_k := \min \{ x_k \text{ s.t. } x \in [a, b], h_i(x) \leq 0 \text{ for all } i \}$$
$$\tilde{b}_k := \max \{ x_k \text{ s.t. } x \in [a, b], h_i(x) \leq 0 \text{ for all } i \}$$

and define the vectors $\tilde{a} := (\tilde{a}_1, \ldots, \tilde{a}_d)$ and $\tilde{b} := (\tilde{b}_1, \ldots, \tilde{b}_d)$, which define the rectangle $\tilde{R} := [\tilde{a}, \tilde{b}]$. As a result we have $\tilde{R} \subset R$ and clearly $\tilde{R}$ contains all the feasible points from $R$. In the Branch and Bound algorithm we will therefore replace the rectangle $R$ by the potentially smaller rectangle $\tilde{R}$.

A crucial element for computing lower bounds on a subset $R$ were the minimum and maximum distances $d^{iR}$ and $D^{iR}$, respectively, between sample points $x^i$ and the convex set $R$. These numbers were easy to compute for rectangles. Now we consider the convex set $C := \tilde{R} \cap \left\{ x \in \mathbb{R}^d : h_i(x) \leq 0 \text{ for } i = 1, \ldots, m \right\}$ and therefore the computation of the numbers may differ. In Algorithm 5.2 for scaling down a rectangle $R$, we will compute the numbers

$$d^{iC} := \min \left\{ \|x^i - x\| \mid x \in \tilde{R}, h_i(x) \leq 0 \text{ for all } i \right\} \tag{5.3}$$

for every sample point $x^i$. Notice that (5.3) is a convex optimization problem and the distance $d^{iC}$ measures the exact (minimal) distance between $x^i$ and $C$. Furthermore, we will compute the numbers

$$D^{i\tilde{R}} := \max \left\{ \|x^i - x\| \mid x \in \tilde{R} \right\}$$

79

for every sample point $x^i$, which potentially overestimates the real (maximum) distance between $x^i$ and $C$ due to $C \subset \tilde{R}$. We do not compute the exakt maximum distance, because this is a nonconvex problem that maybe takes too long to be solved. For the computation of lower bounds the adjusted numbers $d^{iC}, i = 1, \ldots, n$ and $D^{i\tilde{R}}, i = 1, \ldots, n$ will be used instead of the original numbers $d^{iR}$ and $D^{iR}$, see Algorithm 5.2.

---

**Algorithm 5.2:** ScalRect

$\quad$ **Input:** $a, b, h_1(x), \ldots, h_m(x), x^1, \ldots, x^n$

$\quad$ **Output:** $a, b, d^{iR}, D^{iR}$

$\quad$ **for** $k = 1, \ldots, d$ **do**

$\quad\quad$ Compute

$\quad\quad\quad$ $\tilde{a}_k := \min \{x_k | \, x \in [a, b], h_i(x) \leq 0 \text{ for all } i\}$

$\quad\quad\quad$ $\tilde{b}_k := \max \{x_k | \, x \in [a, b], h_i(x) \leq 0 \text{ for all } i\}$

$\quad$ **end**

$\quad$ Set

$\quad\quad$ $\tilde{a} := (\tilde{a}_1, \ldots, \tilde{a}_d)$, $\tilde{b} := (\tilde{b}_1, \ldots, \tilde{b}_d)$ and $\tilde{R} := [\tilde{a}, \tilde{b}]$;

$\quad$ **for** $i = 1, \ldots, n$ **do**

$\quad\quad$ Compute

$\quad\quad\quad$ $d^{iC} := \min \left\{ \|x^i - x\| \, \big| \, x \in \tilde{R}, h_1(x), \ldots, h_m(x) \leq 0 \right\}$

$\quad\quad\quad$ $D^{i\tilde{R}} := \max \left\{ \|x^i - x\| \, \big| \, x \in \tilde{R} \right\}$

$\quad$ **end**

$\quad$ Set

$\quad\quad$ $a \leftarrow \tilde{a}$;

$\quad\quad$ $b \leftarrow \tilde{b}$;

$\quad\quad$ $d^{iR} \leftarrow d^{iC}$ for all $i = 1, \ldots, n$;

$\quad\quad$ $D^{iR} \leftarrow D^{i\tilde{R}}$ for all $i = 1, \ldots, n$.

---

In Section 5.2 we finally present the Branch and Bound algorithm for problem (5.2).


## 5.2 The modified Branch and Bound algorithm

We use Algorithm 3.1 as a basis and emphasize the parts where the new algorithm differs. The modified pseudo code is presented in Algorithm 5.3.

**Algorithm 5.3:** Branch and Bound with convex constraints

---

**Input:** $x^L$ and $x^U$, convex functions $h_1(x), \ldots, h_m(x)$, $x^1, \ldots, x^n$, $\phi$

cheap objective function $g : \mathbb{R}^d \to \mathbb{R}$,

estimation function $\Lambda_g(R, u)_{u \in V} : \mathbb{R}^d_+ \times \mathbb{R}^n_+ \to \mathbb{R}$ (see Rem. 3.11, 3.12 for $V$).

**Output:** Optimum $v$ and optimal objective function value $g(v)$.

**Initialization:**

Set $a \leftarrow x^L$, $b \leftarrow x^U$, and $Rec := R_0 := [a, b]$.

execute ValRect$(R, h_1(x), \ldots, h_m(x))$.

  If output(ValRect) = invalid: break (no solution); else: continue

execute ScalRect$(R, a, b, h_1(x), \ldots, h_m(x), x^1, \ldots, x^n)$.

  Set $R, a, b, d^{iR}, D^{iR} \leftarrow$ output(ScalRect).

Define $V_0$ as in Remark 3.11 resp. 3.12.

Set ub $\leftarrow \min\{g(a), g(b)\}$, $v \leftarrow \operatorname{argmin}\{g(a), g(b)\}$, $N \leftarrow$ Number of iterations for IHR, and execute IHR$(v, \text{ub}, R_0, g, N)$.

Set $v \leftarrow$ output1(IHR), ub $\leftarrow$ output2(IHR), $\mathcal{B} \leftarrow \{Rec\}$, and

lb $\leftarrow$ lb$(Rec) = \Lambda_g(Rec, u)(= \Lambda_g(R_0, u))$ with $u \in V$, stop$\leftarrow$*false*.

**while** *stop=false* **do**

  **if** ub = lb **then**

    |  *stop* $\leftarrow$ *true* ($v$ is optimal solution and ub the opt. obj. function value)

  **else**

    Compute $b_j - a_j = \max\{b_i - a_i : i = 1, \ldots, d\}$ and set

      $a^1 \leftarrow a, b^1 \leftarrow (b_1, \ldots, b_{j-1}, (b_j + a_j)/2, b_{j+1}, \ldots, b_d)^T$

      $a^2 \leftarrow (a_1, \ldots, a_{j-1}, (b_j + a_j)/2, a_{j+1}, \ldots, a_d)^T, b^2 \leftarrow b$

      $R_1 \leftarrow \{x : a^1 \le x \le b^1\}, R_2 \leftarrow \{x : a^2 \le x \le b^2\}.$

    execute ValRect$(R_t, h_1(x), \ldots, h_m(x))$ for $t = 1, 2$

      If output(ValRect) = invalid: delete rectangle (max. 1);

    execute ScalRect$(R_t, a^t, b^t, h_1(x), \ldots, h_m(x), x^1, \ldots, x^n)$ for $t = 1, 2$

      Set $R_t, a^t, b^t, d^{iR_t}, D^{iR_t} \leftarrow$ output(ScalRect) for $t = 1, 2$ and all $i$

    Define $V_1$ and $V_2$ as in Remark 3.11 resp. Remark 3.12. Set

      lb$(R_1) \leftarrow \max\{\text{lb}(R), \Lambda_g(R_1, u)$ with $u \in V_1\}$;

      lb$(R_2) \leftarrow \max\{\text{lb}(R), \Lambda_g(R_2, u)$ with $u \in V_2\}$;

      ub $\leftarrow \min\{\text{ub}, g(b^1), g(a^2)\} = \min\{g(v), g(b^1), g(a^2)\}$;

      $v \leftarrow \arg(\text{ub})$.

    Execute IHR$(v, \text{ub}, R_0, g, N)$ and update

      $v \leftarrow$ IHR-Output(component1); ub $\leftarrow$ IHR-Output(component2);

      $\mathcal{B} \leftarrow \{R \in (\mathcal{B} \setminus \{Rec\} \cup \{R_1, R_2\})$ with lb$(R) < \text{ub}\}$;

      lb $\leftarrow$ $\begin{cases} \min\{\text{lb}(R) : R \in \mathcal{B}\} & \text{if } \mathcal{B} \ne \emptyset \\ \text{ub} & \text{if } \mathcal{B} = \emptyset. \end{cases}$

    Choose $Rec \in \mathcal{B}$ satisfying lb$(Rec) = $ lb. Update $a$ and $b$ such that

    $Rec = [a, b]$.

  **end**

**end**

81

# Chapter 6

# Numerical tests

Before we run and test Algorithm 3.1 respective Algorithm 5.3, which are involved in the RBF-method 2.1, we have to discuss some practical issues concerning the implementation. For example, we have to decide how many starting points we use and where we place them. The minimal number of starting points is theoretically funded (see Theorem 2.6 and 2.7), but not the location. Further questions are which of the three subproblems (minimizing the *interpolating function* $s_n$ from Remark 3.2, solving the *auxiliary problem* from Definition 3.3 or solving the *weighted auxiliary problem* from Definition 3.6) we focuss on in every iteration, i.e. how to choose the weighing parameter $f_n^*$, and how we solve the convex subproblems we discussed in Section 3.4. After talking about these practical issues, we will explain how the program can be used. We finish the chapter with results of our numerical tests.

## 6.1 The implementation

### 6.1.1 Initialisation

An important input for the RBF-method, and therefore for Algorithms 3.1 and 5.3, are the $N$ starting points where the expensive objective is sampled at the beginning. The choice of those starting points has to consider two aspects: The number and the location. As the set of starting points has to be a *unisolvent set* (see Theorem 2.6) the minimal number of starting points depends on the dimension of the problem and on the chosen radial basis function. In Table 6.1 we listed the minimal number of starting points the user has to choose in the program. However, we are nearly free in the choice of the location of the starting points, except that the points have to form a *unisolvent set* (Theorems 2.6 and 2.7). This means e.g. in the case of a two-dimensional problem which we are going to solve with a *cubic spline* or a *thin plate spline*, that the starting points must not be collinear [37].

Table 6.1: Minimal number of starting points to be chosen in the program.

| $d$ | Gaussian, inverse multiquadric, multiquadric, linear spline | thin plate spline, cubic spline |
|---|---|---|
| 1 | 2 | 2 |
| 2 | 2 | 3 |
| 3 | 2 | 4 |
| 4 | 2 | 5 |
| 5 | 2 | 6 |
| 6 | 2 | 7 |

Furthermore, for global optimization it is important to cover the whole feasible set with sample points. To achieve both, distribute the points over the box and to prevent that the sample points lie collinear, we use the "Latin Hypercube sampling" (see e.g. [30]). Here the given box is divided into $N^d$ equal boxes. Now the aim is to choose those $N$ boxes whose midpoint will then be taken as a sample point. We select those boxes as follows: As we have $N$ equidistant intervals in each dimension-direction, we choose, dimension for dimension, one of these intervals randomly and create the first sample point. As we want to prevent to use the same intervals for the next sample point, we "delete" the intervals we have already used. Here "delete" means: We do not take the already used intervals into account when we choose the next intervals randomly to create the following sample point. If we iterate this procedure, we have no choice for the last sample point but have to use the interval that remains. The pseudocode for this procedure, which is implemented in both Algorithms 3.1 and 5.3, is shown in Algorithm 6.1.

Due to the fact that the result of a run depends on the chosen starting points, we use, seperated for each test instance, the same pattern of starting points for the comparing tests.

## 6.1.2 Stopping criteria

We have to determine two stopping criteria both for the Algorithms 3.1 resp. 5.3, which minimize the functions $a_n$, $w_n$ and $s_n$, and for the whole RBF-Method in

---

**Algorithm 6.1:** Latin Hypercube, Selection

---

**Input**: Vectors $x^L$ and $x^U$, Number of sample points $N$, dimension $d$.

**Output**: Sample points $m^1, \ldots, m^N$.

**Initialization:**

Create lists of intervals for each dimension:

$\quad \mathcal{L}^j = \{I_1, \ldots, I_N\}$ for all $j = 1, \ldots, d$.

**for** $k = 1, \ldots, N$ **do**

$\quad$ **for** $j = 1, \ldots, d$ **do**

$\quad\quad$ Choose a random interval $I_i$ uniformly distributed over $\mathcal{L}^j$:

$$I_i = \mathrm{rand}(\mathcal{L}^j)$$

$\quad\quad$ Compute the $j$-th coordinate of the $k$-th sample point:

$$m_j^k = x_j^L - \frac{x_j^U - x_j^L}{2N} + i \cdot \left( \frac{x_j^U - x_j^L}{N} \right)$$

$\quad\quad$ Update $\mathcal{L}^j$:

$\quad\quad\quad \mathcal{L}^j \leftarrow \mathcal{L}^j \backslash \{I_i\}$

$\quad$ **end**

**end**

---

Algorithm 2.1, which minimizes the expensive objective. We start with the first one: In both Algorithms 3.1 and 5.3, the formal stopping criterion is the request "$lb = ub$". For practical issues this criterion is too sharp as within huge computations there usually appear numerical inaccuracies. One idea might be to allow a small gap $\varepsilon$ between $lb$ and $ub$, i.e. to stop the Algorithm when we have "$ub - lb < \varepsilon$" and to tolerate small inaccuracies. The drawback with this stopping criterion is that it is absolute and does not take into account the shape of the function to be minimized. For our implementation we chose the relative stopping criterion

$$ub - lb < \varepsilon \cdot (1 + |ub| + |lb|).$$

This means that for functions with bigger absolute function values the accepted gap is bigger than for functions with smaller absolute function values.

The stopping criterion for the RBF-Method in Algorithm 2.1 depends on the choice of the minimization problem. If we chose one of the testproblems, where the global minimum value `Global Min` is known, we use a relative stopping criterion. This criterion considers the smallest function value $f_{min} = \min\{f(x_1), \dots, f(x_n)\}$ computed so far and looks like

$$\frac{f_{min} - \text{Global Min}}{\text{Global Min}} < 0.01.$$

If we choose an expensive function without a known minimum value we stop the Algorithm after a prescribed number of evaluations `N`.

### 6.1.3 Cycles

We are almost completely free in the choice of which of the cheap objective functions $a_n$ (auxiliary problem), $w_n$ (weighted auxiliary problem) or $s_n$, the interpolating function itself, we want to minimize with Algorithm 3.1 respective Algorithm 5.3 in each iteration step of the RBF-method. Theoretically we have to solve the *auxiliary problem* in Definition 3.3 infinitely often to maintain convergence in the cases where convergence is proved (see [14]). From a practical point of view the aim is to get good function values within only a few iterations, i.e. with only a few sample points. Furthermore, the results in [14] show that it is worthwhile to use cycles which skip the minimizing of $a_n$.

For global optimization local search is typically used alternately with global search.

Since global search can be identified with minimizing $a_n$ and local search with minimizing the interpolating function $s_n$, the third function $w_n$, which is a mixed form of both, can be used to lead stepwise from global to local search. Here the parameter $f^*$ regulates how both functions $a_n$ and $s_n$ are weighted. Gutmann [14] tested several cycles and suggested a length of at most 4. The cycles typically start with either minimizing $a_n$ or $w_n$ with a very big $|f^*|$, i.e. global search, and end with minimizing $s_n$, i.e. local search. Between those two extremes the function $w_n$ is minimized, with $|f^*|$ chosen very big in the beginning and then decreased step by step, emphazising more and more the local search. For our numerical tests we chose a cycle that takes small function values into account (see [14]). This cycle is constructed as follows: In each iteration $n$ of the RBF-Method let $f(x^1), \ldots, f(x^n)$ be the function evaluations made so far, which we sort in increasing order: $f(x_{\sigma(1)}) \leq \ldots \leq f(x_{\sigma(n)})$, where $\sigma$ is the associated permutation of $\{1, \ldots, n\}$. We define $a := \mathrm{mod}(n, 3)$. With

$$K_n := \left\lfloor \frac{3-a}{3} \cdot (n+1) \right\rfloor$$

we regulate how many of the small function values we take into account. To keep the notation clear, we abbreviate the minimum of the current $s_n$ with "$s_{min}$" and write "$f_n^* = -\infty$" if we want to minimize $a_n$ (see the last cycle) and "$f_n^* = s_{min}$" if we want to minimize $s_n$. With these parameters $a$ and $K_n$ we can formulate the cycle

$$f_n^* = \begin{cases} s_{min} - \left(\frac{3-a}{3}\right)^2 \cdot \left(\max_{1 \leq i \leq K_n} f(x_{\sigma(i)}) - s_{min}\right) & \text{, if } a \in \{1, 2\} \\ s_{min} & \text{, if } a = 0 \end{cases}$$

which fulfills the requirement of leading from global to local search strategies as discussed.

Within this cycle the choice $f^* = s_{min}$ potentially leads to the computation of a sample point that equals one of the already made sample points $\{x^1, \ldots, x^n\}$. To use twice the same sample point would lead to a singular matrix $A$ (see Theorem 2.7) and would keep the Algorithm from progressing. In the beginning of writing the code we handled this problem by computing the distance of the computed potential sample point to the points $x^1, \ldots, x^n$ and discarding this new point if this distance is smaller than a prescribed value. As this fixed prescribed value could prevent the algorithm from reaching a global optimum, we abandoned this strategy. Instead we followed the idea [14], to compare the minimal value $s_{min}$ with $f_{min} = \min\{f(x^1), \ldots, f(x^n)\}$, and to discard the potential new sample point $\arg s_{min}$ if the quotient $\frac{f_{min} - s_{min}}{|f_{min}|}$ is smaller than $10^{-4}$.

Furthermore, we use some naive cycles for comparison. These are firstly to minimize the interpolating function $s_n$ in every iteration step, i.e. to set

$$f_n^* = s_{min} \quad \text{for all} \quad n$$

which can be seen as a pure local search strategy.

As a second naive cycle, we minimize the function $a_n$ in every iteration step, i.e. we set

$$f_n^* = -\infty \quad \text{for all} \quad n$$

which stands for pure global search.

### 6.1.4 Solving the convex subproblems

To compute the lower bounds proposed in Section 3.4 we have to solve a lot of simple convex subproblems. For these subproblems we use the software IPopt of the open-source community CoinOR (http://www.coin-or.org/). As the dimension of these subproblems is equal to the number of already sampled points, the dimension of the subproblems grows in every iteration step where a new sample is made. Within the numerical tests we saw that the runtime to solve these subproblems grew rapidly and made the progressing of the whole algorithm much slower. To make the reader understand the underlying problem we will show in an example (see Table 6.2) how the runtime increases with every iteration step.

### 6.1.5 Sorting of boxes

Within the algorithms a huge number of rectangles is stored and repeatedly subdivided. To keep the algorithm from creating disproportionally small rectangles, mainly in the beginning, we enable the user to sort rectangles with a diameter smaller than a preassumed value into a separated list. We tested different values for the feasible diameter.

## 6.2 How to use the code

### 6.2.1 Parameters for the input

The program is written in $C\#$. With the start of the program a GUI will be opened, where different parameters have to be chosen by the user.

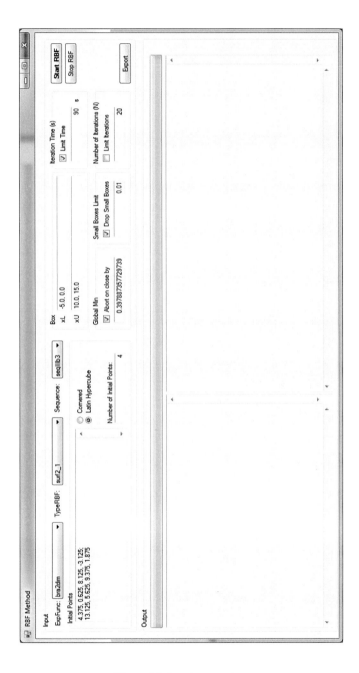

Figure 6.1: Input parameters

Within the form in Figure 6.1 the different inputs are:

| | | |
|---|---|---|
| `ExpFunc` | - | expensive function (test problem) |
| `TypeRBF` | - | radial basis function |
| `Sequence` | - | choice of the cycle (see Section 6.1.3; in addition some other cycles are implemented) |
| `Number of Initial Points` | - | number of starting points (see Table 6.1) |
| `Box(xL)` | - | lower left vertex of the box $[x^L, x^U]$ |
| `Box(xU)` | - | upper right vertex of the box $[x^L, x^U]$ |

Furthermore the user can choose how the starting points are created. With one of the two alternatives

| | | |
|---|---|---|
| `Cornered` | - | vertices of the `Box` |
| `Latin Hypercube` | - | Latin Hypercube sample (see Section 6.1) |

the form is automatically filled with the `Initial Points`.

Moreover it is possible to allow or disallow the use of a global minimum value for a stopping criterion, to limit the number of iterations and to decide whether small boxes are sorted in a different list:

| | | |
|---|---|---|
| `Global Min` | - | use global minimum yes /no, value |
| `Number of Iterations` | - | limit the number yes /no, integer number |
| `Small Boxes Limit` | - | sorting off small boxes yes /no, diameter |

If the user runs the program with one of the test problems he/she can choose this test problem in `ExpFunc`. With the choice of the expensive function `ExpFunc` the values `Box(xL)`,`Box(xU)` and `Global Min` are automatically written. It also possible to run the algorithm without the specification of a minimal value.

With `TypeRBF` the user chooses between the following radial basis functions (see also Definition 2.1):

| | | | |
|---|---|---|---|
| `surf1_0` | - | $\phi(r) = r$ | (linear spline) |
| `surf2_1` | - | $\phi(r) = r^2 \log r$ | (thin plate spline) |
| `surf3_1` | - | $\phi(r) = r^3$ | (cubic spline) |
| `gauss1_neg1` | - | $\phi(r) = \exp(-r^2)$ | (Gaussian) |
| `multiquad05_0` | - | $\phi(r) = \sqrt{r^2 + 1}$ | (multiquadric) |
| `invmultiquad05_neg1` | - | $\phi(r) = (\sqrt{r^2 + 1})^{-1}$ | (inverse multiquadric) |

With `Sequence` the user chooses the cycle, i.e. the order in which the functions $a_n$, $s_n$ and $w_n$ are minimized (see Section 6.1.3).

With the `Start RBF`-button the user starts the RBF-Method.

### 6.2.2   Reading the output

The output we get from the code contains different information concerning the bounds, the sample points, the runtime etc. In Figure 6.2 we see an example of such an output.

We can see that the output is divided in two parts: The left part consists of current changing numbers like number of boxes, runtime etc. The right part mainly shows the produced sample points and their expensive function values. We first give some more detailed explanations for the left hand side: In the first block we see the current subproblem that is solved and the value of $f^*$ (in case we are minimizing $w_n$). This is followed by an information block of the box subdivisions. Here we can see how many iterations (which is equal to the number of boxdivisions) the algorithm which minimizes one of the three functions $s_n$, $a_n$ and $w_n$, has done, how many boxes are very small and how many boxes are deleted. With "small" we mean that the diameter of a box is smaller than a prescribed value, which can be chosen by the user as explained in Section 6.1. We collect these small boxes in a separate list and do not consider them for further divisions, but only store their lower bounds.

In the next block of information we see the current global upper and lower bounds with the two upper bounds and two lower bounds of the last divided box. Here we can see how big the current gap between the global lower and upper bounds is.

The right hand side of the output can be read as follows: Here each row presents one iteration step of the RBF-Method, i.e. each row contains the information for exactly one sample point of the expensive objective. The first entry is the function value itself followed by the corresponding sample point. The numbers in the last columns are the final time used for the computation of this single iteration step and the final number of box divisions done for this single iteration step. In the case of the problems with additional convex constraints (see Example 1-4 from Section 6.3.4) we added in addition the value of the constraint violation. Here a negative sign means that the sample point is feasible.

## 6.3   Test instances and runtime

We tested both box constrained test problems from the literature (see Appendix) and problems with additional convex constraints. We saw that the runtime grows rapidly with the number of sample points. Therefore, we were first of all interested to find out what causes this enormous time consumption. We tested how the runtime is

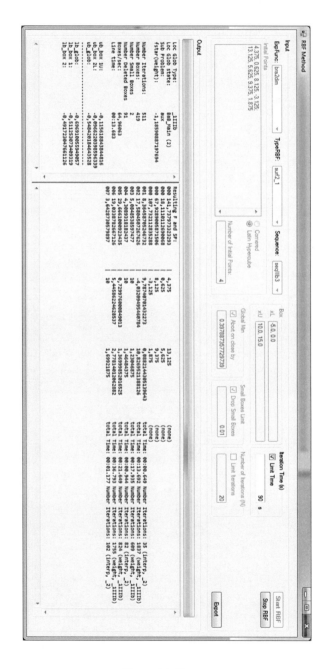

Figure 6.2: Output

92

distributed on the several subproblems that have to be solved. In addition we tested alternative lower bounds. Furthermore we were interested in how the algorithm succeeds when we use different values for the small box limits resp. when we restrict the time for solving each of the subproblems $s_n$, $w_n$ or $a_n$. For all the latter tests, which should give us information about the time consumption, we always use the same test problem "cam" (see Appendix).

Furthermore, to distinguish the iterations of the RBF-Method from the iterations of the subsolver, we will call the iterations of the RBF-Method "iteration steps".

## 6.3.1  Runtime

*Runtime subsolver:*
For the testproblem "cam" we run Algorithm 3.1 19 iteration steps, i.e. we solved 19 subproblems of the type $s$ or $w$. In Table 6.2 we distinguish between the total time for solving one subproblem, the percentage of total time that is used for solving the convex subproblems and the percentage that is used for the IHR. In addition we list the number of total iterations, i.e. box divisions, and the average time that is used for solving one convex subproblem. We can see that in the beginning the algorithm needs only a few seconds for one iteration step, i.e. to produce one row in the output, and that the subsolver only needs some milliseconds to solve one convex subproblem. This time grows rapidly as the algorithm progresses and in the last iteration steps the subsolver needs more than five seconds for one convex subproblem. In addition the number of iterations, i.e. of box divisions, grew rapidly from nine to several thousand. Both leads to the fact that the algorithm needs more than two minutes each for iterations 17 and 18. In addition Table 6.2 shows that the subsolver uses up to 95% of the total time needed for one iteration step. The IHR-routine uses a much smaller percentage of the total runtime - except for the $s_n$, where the total runtime is only a few seconds.

*Using lower bound $\Lambda_s$ for $s_n$ (see Corrolaries 3.14 and 3.15):*
In [14] lower bounds for $s_n$ were computed using linear underestimators inspired by [26]. We implemented both strategies in our RBF-Method and compared their performance. In Table 6.3 we see the results of ten iteration steps with the expensive function "cam" and the inverse multiquadric radial basis function. The time used for each iteration was limited to 90 seconds.

Table 6.3 shows that the computation of the lower bounds from [14] is much faster than the computation of our lower bounds $\Lambda_s$ for this example, and we noted that our

Table 6.2: Runtime subsolver, test problem "cam"

| iter step $n$ | $f(x)$ | type sub | time tot min:sec | % sub | % IHR | iter | av msec/sub |
|---|---|---|---|---|---|---|---|
| 01 | $-0.74$ | $s_1$ | 00:00.37 | 0 | 69.62 | 9 | 0 |
| 02 | 0.79 | $w_2$ | 00:06.33 | 4.08 | 14.65 | 33 | 7 |
| 03 | $-0.17$ | $w_3$ | 00:06.73 | 57.93 | 38.44 | 565 | 3 |
| 04 | $-0.21$ | $s_4$ | 00:00.72 | 0 | 81.56 | 49 | 0 |
| 05 | 0.62 | $w_5$ | 00:06.33 | 81.44 | 16.35 | 536 | 13 |
| 06 | $-0.64$ | $w_6$ | 01:01.03 | 90.55 | 8.11 | 1483 | 19 |
| 07 | $-0.33$ | $s_7$ | 00:00.97 | 0 | 74.87 | 87 | 0 |
| 08 | $-0.45$ | $w_8$ | 01:34.05 | 95.64 | 3.27 | 532 | 179 |
| 09 | $-0.28$ | $w_9$ | 01:34.42 | 95.87 | 3.05 | 426 | 302 |
| 10 | $-0.65$ | $s_{10}$ | 00:02.36 | 0 | 64.58 | 235 | 0 |
| 11 | 0.37 | $w_{11}$ | 01:36.73 | 95.77 | 3.04 | 334 | 935 |
| 12 | $-0.45$ | $w_{12}$ | 01:41.46 | 93.14 | 4.46 | 619 | 1852 |
| 13 | $-0.99$ | $s_{13}$ | 00:06.86 | 0 | 58.38 | 662 | 0 |
| 14 | $-0.58$ | $w_{14}$ | 01:45.55 | 92.35 | 4.87 | 647 | 3361 |
| 15 | $-0.36$ | $w_{15}$ | 01:42.96 | 92.05 | 5.05 | 614 | 3791 |
| 16 | $-0.99$ | $s_{16}$ | 00:07.48 | 0 | 57.57 | 615 | 0 |
| 17 | $-0.34$ | $w_{17}$ | 02:05.14 | 82.32 | 10.24 | 1526 | 4906 |
| 18 | $-0.66$ | $w_{18}$ | 02:24.10 | 69.67 | 16.84 | 3369 | 5284 |
| 19 | $-0.99$ | $s_{19}$ | 00:51.52 | 0 | 53.80 | 3932 | 0 |
| opt | $-1.03$ | | | | | | |

tests in general run faster with those lower bounds. Thus, in the case of computing lower bounds for $s_n$, we followed [14].

*Using lower bound $\Lambda_a^1$ resp. $\Lambda_a^2$ for $a_n$ (see Corollary 3.32):*
In Corollar 3.32 we proposed two alternative lower bounds, $\Lambda_a^1$ and $\Lambda_a^2$, that can be used within the Branch and Bound-routine for minimizing $a_n$. We suspected $\Lambda_a^1$ would perform worse than $\Lambda_a^2$, because its estimation was very rough. We compared both lower bounds and ran a test with the testfunction "cam" and the inverse multiquadric radial basis function. The total runtime per iteration step was limited to 90 seconds.

Table 6.3: Comparison of the computation time for lower bounds for $s_n$.

| iter step $n$ | type sub | lower bound $\Lambda_s$ | | lower bound from [14] | |
|---|---|---|---|---|---|
| | | time tot min:sec | iter | time tot min:sec | iter |
| 01 | $s_1$ | 00:09.70 | 87 | 00:00.30 | 5 |
| 02 | $s_2$ | 01:30.20 | 28484 | 00:00.40 | 44 |
| 03 | $s_3$ | 01:30.17 | 29370 | 00:00.49 | 63 |
| 04 | $s_4$ | 01:30.27 | 28168 | 00:00.93 | 142 |
| 05 | $s_5$ | 01:30.30 | 26616 | 00:01.82 | 277 |
| 06 | $s_6$ | 01:30.26 | 26352 | 00:02.18 | 278 |
| 07 | $s_7$ | 01:30.33 | 26251 | 00:02.43 | 292 |
| 08 | $s_8$ | 01:30.39 | 25997 | 00:05.77 | 798 |
| 09 | $s_9$ | 01:30.35 | 26165 | 00:06.97 | 883 |
| 10 | $s_{10}$ | 01:30.35 | 26115 | 00:23.49 | 2955 |

Table 6.4: Comparison of the computation time of the lower bounds $\Lambda_a^1$ and $\Lambda_a^2$.

| iter step $n$ | type sub | lower bound $\Lambda_a^1$ | | lower bound $\Lambda_a^2$ | |
|---|---|---|---|---|---|
| | | time tot min:sec | iter | time tot min:sec | iter |
| 01 | $a_1$ | 00:34.70 | 12510 | 00:00.84 | 40 |
| 02 | $a_2$ | 01:30.18 | 27318 | 00:00.61 | 29 |
| 03 | $a_3$ | 01:30.17 | 26944 | 00:00.43 | 20 |
| 04 | $a_4$ | 01:30.18 | 26318 | 00:03.55 | 246 |
| 05 | $a_5$ | 01:30.21 | 25415 | 00:03.68 | 248 |
| 06 | $a_6$ | 01:30.20 | 25226 | 00:04.99 | 320 |
| 07 | $a_7$ | 01:30.19 | 23993 | 00:09.83 | 639 |
| 08 | $a_8$ | 01:30.24 | 22441 | 00:16.01 | 1042 |
| 09 | $a_9$ | 01:30.25 | 21494 | 00:17.85 | 1128 |
| 10 | $a_{10}$ | 01:30.23 | 20439 | 00:21.49 | 1348 |

Table 6.4 shows that our conjecture was right and that $\Lambda_a^2$ works much better than $\Lambda_a^1$. The allowed runtime for one iteration step was comletely exploited when we used $\Lambda_a^1$. It seems that not even the stopping criterion was reached, except for the first iteration step. This shows that the lower bounds $\Lambda_a^1$ are of worse quality compared to the lower bounds $\Lambda_a^2$. The number of box divisions necessary for Branch and Bound with $\Lambda_a^1$ is, for every iteration, much larger than in the case of using $\Lambda_a^2$. Table 6.4 shows in addition that the number of box divisions definitely grows in every iteration

step.

As the runtime of the RBF-Method grows a lot in every iteration step, we tried to find tools to limit the runtime: A successful method was to discard small boxes in order to limit the number of box divisions in each iteration step, limiting the time per iteration step.

*Selecting small boxes separated:*

To prevent the RBF-Method from searching in unproportionally small boxes, we include the tool of limiting the number of boxes that are divided by discarding those boxes which have a diameter smaller than a prescribed value. Of course this strategy might prevent the algorithm from finding the global optimum. To test the effect of this measure we ran the RBF-Method 20 iteration steps with the testfunction "cam" three times - each with a different radial basis function. For each of the 20 iteration steps we limited the time to 90 seconds. We compared the performance of the algorithm: In the first case we discarded boxes with a diameter smaller than 0.01 and compared the results to the case without diameter restriction.

Table 6.5: Diameter restriction yes /no

|  | with diameter restriction | | | without diameter restriction | | |
|---|---|---|---|---|---|---|
|  | inverse multiquadric | thin plate spline | cubic spline | inverse multiquadric | thin plate spline | cubic spline |
| best value | 0.2415 | −0.6751 | −0.7694 | −0.4820 | −0.9399 | −0.9643 |
| runtime (minutes) | 9.85 | 8.38 | 33.91 | 9.25 | 11.97 | 41.43 |

Table 6.5 shows that, as expected, the best function value ("best value") that is produced in each of the three cases is much better when we skip the diameter-restriction. But we see in addition that, except for the inverse multiquadric radial basis function, the runtime grows.

Next we compared the performance of the three different cases when we restrict the diameter to 0.01 respectivly to 0.001. We see in Table 6.6 that the runtime grows in the case of 0.001 compared to the case 0.01, but that for the thin plate spline radial basis function the best function value ("best value") that is produced is better than in the case of no restriction. So to restrict the diameter of the boxes does not necessarily lead to worse performance.

*Time limit:*

In a next step we compared how the performance changes when we limit the time per

Table 6.6: Diameter restriction, different diameters

| | diameter restriction 0.01 | | | diameter restriction 0.001 | | |
|---|---|---|---|---|---|---|
| | inverse multiquadric | thin plate spline | cubic spline | inverse multiquadric | thin plate spline | cubic spline |
| best value | 0.2415 | −0.6751 | −0.7694 | −0.0991 | −1.0161 | −0.9116 |
| time (minutes) | 9.85 | 8.38 | 33.91 | 12.38 | 27.66 | 53.26 |

iteration. We run the same example as above for 20 iteration steps with a diamter restriction of 0.01. First we restricted the time per iteration to 90 seconds and second we skipped the time restriction.

Table 6.7: Timelimit yes / no

| | time limit 90 sec | | | without time limit | | |
|---|---|---|---|---|---|---|
| | inverse multiquadric | thin plate spline | cubic spline | inverse multiquadric | thin plate spline | cubic spline |
| best value | 0.2415 | −0.6751 | −0.7694 | −0.8856 | −1.0255 | −0.9784 |
| time (minutes) | 9.85 | 8.38 | 33.91 | 14.85 | 11.06 | 137.06 |

Table 6.7 shows that of course the best values are improved if there is no time limit. In addition we see that in the case of the cubic spline radial basis function the runtime grows rapidly, whereas the runtime using the thin plate spline radial basis function only grows moderately.

The previous comparisons lead to the result that both, time limits and the restriction of diamters, worsen the performance, i.e. reduce the quality of the best values. Therefore it might be desirable to use none of these time saving strategies. In a last step, we tried to run two of the three cases without a timelimit and without any restriction of diameters. The results can be seen in Table 6.8.

In the case where we used the inverse multiquadric radial basis function we needed 45 minutes and the result is not very close to the optimal value −1.03. In the case where we used the thin plate spline radial basis function we needed more than four hours to compute eleven iteration steps and the best value is much worse than in

Table 6.8: 20 iteration steps, without timelimit, without diameter restriction

|  | inverse multiquadric | thin plate spline |
|---|---|---|
| best value | $-0.6952$ | 0.13 (11 iter steps) |
| time (minutes) | 45.84 | > 240 |

the first case.

As a consequence we can say that we have to use and balance tools like time or diameter restrictions to compute usable results.

## 6.3.2 Box constrained problems

We tested the RBF-Method with several global test instances from the literature, which are listed in the Appendix.

## 6.3.3 Results: Box constrained problems

In all tests with the box constrained problems we ran Algorithm 2.1 30 iteration steps, limited each iteration time to three minutes and refrained from subdividing boxes with a diameter smaller than 0.01. We tested all testproblems each with the four radial basis functions thin plate spline, cubic spline, multiquadric and inverse multiquadric. In the table we list the dimension of the problem in the second column and the global optimum of each problem in the third column. Table 6.9 shows the results, i.e. the best expensive function value that was produced within 30 iteration steps. The numbers in parentheses are, in case of convergence, the number of iteration steps that were needed to compute the global minimum.

As Table 6.9 shows, the problem "hm3" was solved in all four different cases of radial basis functions. Furthermore, the thin plate spline radial basis function gave the best results compared to all other tested radial basis functions. In spite of running the RBF-Method only 30 iteration steps, we have some promising results with the thin plate spline radial basis function.

## 6.3.4 Box constrained problems with additional convex constraints

We expanded Algorithm 3.1 such that we can solve problems with additional convex constraints with the RBF-method. In Algorithm 5.3 boxes which are infeasible with

Table 6.9: Boxconstrained problems, 30 iteration steps

| problem | dim | opt | thin plate spline | cubic spline | multi- quadric | inverse multiquadric |
|---------|-----|-----|-------------------|--------------|----------------|----------------------|
| gpr | 2 | 3.0 | 6.1725 | 125.6611 | 12.3949 | 14.3030 |
| brn | 2 | 0.3978 | 0.4507 | 1.9433 | 0.6336 | 2.0377 |
| hm3 | 3 | −3.8627 | −3.8379 (18) | −3.82809 (22) | −3.8353 (23) | −3.8351 (22) |
| sh5 | 4 | −10.1531 | −2.5183 | −1.4993 | −0.5526 | −0.2317 |
| hm6 | 6 | −3.3223 | −2.5808 | −2.5072 | −1.9968 | −2.3247 |

respect to the convex constraints are discarded. We run the algorithm with the following examples.

**Example 1**

The box constrained test problem "brn" has three global optima (see Appendix). With the additional convex constraint

$$(x_1 + \pi)^2 + (x_2 - 12.275)^2 \leq 62.785$$

we cut off the two optima $(9.4247, 2.4749)$ and $(\pi, 2.2750)$ such that the remaining problem

$$\min \left( x_2 - \frac{5.1}{4\pi^2} \cdot x_1^2 + \frac{5}{\pi} \cdot x_1 - 6 \right)^2 + 10 \left( 1 - \frac{1}{8\pi} \right) \cos(x_1) + 10$$

$$s.t. \quad \begin{pmatrix} x_1 \\ x_2 \end{pmatrix} \in \left[ \begin{pmatrix} -5 \\ 0 \end{pmatrix}, \begin{pmatrix} 10 \\ 15 \end{pmatrix} \right]$$

$$(x_1 + \pi)^2 + (x_2 - 12.275)^2 - 62.785 \leq 0$$

has the single global optimum $(-\pi, 12.275)$. It is integrated with the name `bra2dim_conv` in the input-form of the program.

We visualize the problem and the outcome of the RBF-method: Figure 6.3 shows the feasible set and the level curves of the function "brn". In the center of each of the three black ovals which form valleys one of the three global optima of the original box constrained problem is located. We see that with the additional convex condition indeed two of the optima are cut off. Next we run the RBF-method, including Algorithm 5.3, with three fixed starting points, the parameters `surf3_1` (cubic spline radial basis function), `seqIIIb3` (i.e. the first cycle formulated in Section 6.1.3) and using the `Small Boxes Limit` with maximal diameter 0.01. The algorithm terminated after 19 iteration steps finding the global optimum. In Figure 6.3 we see 21 sample points visualized as red crosses, and the last optimal sample point is plotted with a black asterisk.

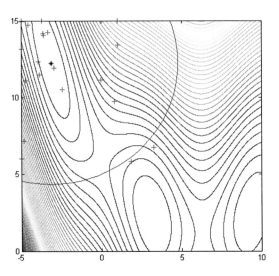

Figure 6.3: 21 iteration steps, convex constraint

We can see that two sample points are outside the feasible set. This comes from the fact that the Branch and Bound algorithm used for one iteration step approximates the feasible set from the outside with rectangles on which the auxiliary functions are minimized.

Further test results for the convex constrained problems are shown in Section 6.3.5.

**Example 2**

The box constrained test problem "cam" has two global optima (see Appendix). With the additional convex constraint

$$(x_1 + 3)^2 + (x_2 - 2)^2 \leq 13$$

100

we cut off the optimum $(0.0898, -0.7126)$ such that the remaining problem

$$\min \left(4 - 2.1x_1^2 + \frac{x_1^4}{3}\right) x_1^2 + x_1 x_2 + \left(-4 + 4x_2^2\right) x_2^2$$

$$s.t. \quad \begin{pmatrix} x_1 \\ x_2 \end{pmatrix} \in \left[\begin{pmatrix} -2 \\ -2 \end{pmatrix}, \begin{pmatrix} 2 \\ 2 \end{pmatrix}\right]$$

$$(x_1 + 3)^2 + (x_2 - 2)^2 - 13 \le 0$$

has the single global optimum $(-0.0898, 0.7126)$.

## Examples 3a and 3b

The box constrained test problem "hm3" has the global optimum $x_{glob} = (0.1, 0.5559, 0.8521)$ (see Appendix). With the additional convex constraints

$$x_1^2 + (x_2 - 1)^2 + (x_3 - 1)^2 \le 1$$

respectively

$$x_1^2 + x_2^2 + x_3^2 \le \|x_{glob}\|^2$$

we cut off a part of the box. With the parameters

$$A = \begin{pmatrix} 3.0 & 10 & 30 \\ 0.1 & 10 & 35 \\ 3.0 & 10 & 30 \\ 0.1 & 10 & 35 \end{pmatrix}, P = \begin{pmatrix} 0.3689 & 0.1170 & 0.2673 \\ 0.4699 & 0.4387 & 0.7470 \\ 0.1091 & 0.8732 & 0.5547 \\ 0.0381 & 0.5743 & 0.8828 \end{pmatrix} \text{ and } c = \begin{pmatrix} 0.1 \\ 1.2 \\ 3.0 \\ 3.2 \end{pmatrix},$$

we create two problems:

Problem 3a

$$\min - \sum_{i=1}^{4} c_i \exp\left(-\sum_{j=1}^{3} A_{ij} (x_j - P_{ij})^2\right)$$

$$s.t. \quad \begin{pmatrix} x_1 \\ x_2 \\ x_3 \end{pmatrix} \in \left[\begin{pmatrix} 0 \\ 0 \\ 0 \end{pmatrix}, \begin{pmatrix} 1 \\ 1 \\ 1 \end{pmatrix}\right]$$

$$x_1^2 + (x_2 - 1)^2 + (x_3 - 1)^2 - 1 \le 0$$

respectively

Problem 3b

$$\min - \sum_{i=1}^{4} c_i \exp\left(-\sum_{j=1}^{3} A_{ij} (x_j - P_{ij})^2\right)$$

$$s.t. \quad \begin{pmatrix} x_1 \\ x_2 \\ x_3 \end{pmatrix} \in \left[\begin{pmatrix} 0 \\ 0 \\ 0 \end{pmatrix}, \begin{pmatrix} 1 \\ 1 \\ 1 \end{pmatrix}\right]$$

$$x_1^2 + x_2^2 + x_3^2 - \|x_{glob}\|^2 \le 0.$$

Both problems 3a and 3b have the single global optimum $x_{glob}$. In problem 3b this optimum lies on the boundary of the feasible set, whereas in problem 3a the optimum lies in the interior.

**Example 4**

The box constrained test problem "sh5" has one global optimum (see Appendix). With the additional convex constraint

$$(x_1 - 4)^2 + (x_2 - 4)^2 + (x_3 - 4)^2 + (x_4 - 4)^2 \leq 8$$

we cut off a part of the box. The remaining problem, with

$$A = \begin{pmatrix} 4 & 4 & 4 & 4 \\ 1 & 1 & 1 & 1 \\ 8 & 8 & 8 & 8 \\ 6 & 6 & 6 & 6 \\ 3 & 7 & 3 & 7 \end{pmatrix} \quad \text{and } c = \begin{pmatrix} 0.1 \\ 0.2 \\ 0.2 \\ 0.4 \\ 0.4 \end{pmatrix},$$

is

$$\min \; -\sum_{i=1}^{5} \left( c_i + \sum_{j=1}^{4} (x_j - A_{ij})^2 \right)^{-1}$$

$$s.t. \quad \begin{pmatrix} x_1 \\ x_2 \\ x_3 \\ x_4 \end{pmatrix} \in \left[ \begin{pmatrix} 0 \\ 0 \\ 0 \\ 0 \end{pmatrix}, \begin{pmatrix} 10 \\ 10 \\ 10 \\ 10 \end{pmatrix} \right]$$

$$(x_1 - 4)^2 + (x_2 - 4)^2 + (x_3 - 4)^2 + (x_4 - 4)^2 - 8 \leq 0.$$

It still has the single global optimum $(4, 4, 4, 4)$.

## 6.3.5   Results: Problems with additional convex constraints

In all tests with the Examples 1-4 we ran the algorithm for 15 iteration steps, limited each iteration step time to three minutes and divided no box whose diameter was smaller than 0.01. In all cases we used the thin plate spline radial basis function.

Table 6.10: Examples 1-4, 15 iteration steps with thin plate spline radial basis function

| problem | opt | best value found | best point found |
|---|---|---|---|
| Example 1 | 0.3978 | 0.4118 | $(-3.0887, 12.1714)$ |
| Example 2 | $-1.0316$ | $-0.6833$ | $(-0.0542, 0.8928)$ |
| Example 3a | $-3.8627$ | $-3.3026$ | $(0.1409, 0.4303, 0.8701)$ |
| Example 3b | $-3.8627$ | $-3.6629$ | $(0.1251, 0.4824, 0.8625)$ |
| Example 4 | $-10.1531$ | $-0.3741$ | $(6.3426, 6.5173, 8.1640, 6.7313)$ |

Table 6.10 shows that in Examples 1 and 3b the best value found is less than 10% worse than the optmal value. Example 3a shows a deviation of less than 15%. Comparing Examples 3a and 3b the algorithm produces a better result in the case where the optimum lies on the boundary of the feasible set.

# Chapter 7

# Conclusions

In the year 2001 the "Radial Basis Function Methods for Global Optimization" were proposed, which seemed to be a very promising surface method for solving global optimization problems. This method had many advantages over other optimization methods, for example, the nearly free choice of starting points and the easy way of implementation. A user of the radial basis function method has a huge liberty of how to use a radial basis function method. He/she can choose an arbitrary radial basis function, a polynomial, the order, in which the subproblems have to be solved etc. But this liberty implied by the same reason many open questions concerning the best choice of all these parameters. Some authors tackled these questions. One question, which was not often discussed, is how the subproblems within the method should be solved. One drawback of these subproblems is that their objective has many local optima, i.e. the original hard global optimization problem leads to a sequence of global optimization problems which are also hard to solve. Since the convergence theorems for the radial basis methods do not depend on the exact computation of the global optimum, heuristics like the DIRECT algorithm are often used to solve these subproblems. Nonetheless the minimization of the subproblems is a very important part of the radial basis function methods.

Furthermore, the easy structure of these subproblems is an advantage which we used to develop an exact method to solve these problems. We put a lot of effort into the computation of sophisticated lower bounds, used the finally developed bounds for a Branch and Bound routine and implemented a program which is able to produce good sample points within a few iterations. In addition, we demonstrated that the use of too simple chosen lower bounds leads to much worse results compared to the results produced by the sophisticated lower bounds. One difficulty within the computation of the bounds was the solution of a huge number of simple convex subproblems within the routine. We found that the solution of these subproblems causes a very long runtime. We overcame this difficulty by incorporating limit tools. We demonstrated that these limit tools have to be chosen in a very balanced way. Although now the detection of a global minimum of the subproblem is no longer guaranteed, the routine produces in parts good results with only very few sample points.

Moreover, we developed a routine to solve not only box constrained problems but also problems with additional arbitrary convex constraints. For black box optimization this task is very difficult. This is why many articles only treat box-contraints (see [7]). The authors of [38] tested their algorithm for a single constrained problem, and beyond that articles in the black box optimization which tackle constrained problems are hard to find. We hope that our method for solving black box problems with convex constraints can be extended for larger problems. As a result of the numerical tests we think that it is worthwhile to put much effort in the computation of exact solutions for the subproblems that arise within the radial basis function methods.

# Appendix A

# Testproblems

Our testproblems were formulated in [9]. We use the following abbreviations:

$x^L$   -   lower left corner of the constraint box
$x^U$   -   upper right corner of the constraint box
glob   -   global minimum (function value)
nglob   -   number of global minima within $[x^L, x^U]$
xglob   -   global minima

**Test problem "Goldstein-Price"** [9], abbr: **gpr**, Dimension: **2**

$$\min \left[1 + (x_1 + x_2 + 1)^2(19 - 14x_1 + 3x_1^2 - 14x_2 + 6x_1x_2 + 3x_2^2)\right]$$
$$\times \left[30 + (2x_1 - 3x_2)^2(18 - 32x_1 + 12x_1^2 + 48x_2 - 36x_1x_2 + 27x_2^2)\right]$$

$$s.t. \quad \begin{pmatrix} x_1 \\ x_2 \end{pmatrix} \in \left[\begin{pmatrix} -2 \\ -2 \end{pmatrix}, \begin{pmatrix} 2 \\ 2 \end{pmatrix}\right]$$

| $x^L$ | [-2; -2] |
|---|---|
| $x^U$ | [2; 2] |
| fglob | 3 |
| nglob | 1 |
| xglob | [0; -1] |

Test problem "Branin"[9], abbr: **brn**, Dimension: **2**

$$\min \left( x_2 - \frac{5.1}{4\pi^2} \cdot x_1^2 + \frac{5}{\pi} \cdot x_1 - 6 \right)^2 + 10 \left( 1 - \frac{1}{8\pi} \right) \cos(x_1) + 10$$

$$s.t. \quad \begin{pmatrix} x_1 \\ x_2 \end{pmatrix} \in \left[ \begin{pmatrix} -5 \\ 0 \end{pmatrix}, \begin{pmatrix} 10 \\ 15 \end{pmatrix} \right]$$

| $x^L$ | [-5; 0] |
|---|---|
| $x^U$ | [10;15] |
| fglob | 0.397887357729739 |
| nglob | 3 |
| xglob | [ 9.42477796;  2.47499998] |
| | [-3.14159265;12.27500000] |
| | [ 3.14159265;  2.27500000] |

Test problem "Six-hump camel"[9], abbr: **cam**, Dimension: **2**

$$\min \left( 4 - 2.1x_1^2 + \frac{x_1^4}{3} \right) x_1^2 + x_1 x_2 + \left( -4 + 4x_2^2 \right) x_2^2$$

$$s.t. \quad \begin{pmatrix} x_1 \\ x_2 \end{pmatrix} \in \left[ \begin{pmatrix} -2 \\ -2 \end{pmatrix}, \begin{pmatrix} 2 \\ 2 \end{pmatrix} \right]$$

| $x^L$ | [-3;-2] |
|---|---|
| $x^U$ | [ 3; 2] |
| fglob | -1.0316284535 |
| nglob | 2 |
| xglob | [ 0.08984201;-0.71265640] |
| | [-0.08984201;  0.71265640] |

Test problem "Shekel 5"[9], abbr: **sh5**, Dimension: **4**

With the parameters

$$A = \begin{pmatrix} 4 & 4 & 4 & 4 \\ 1 & 1 & 1 & 1 \\ 8 & 8 & 8 & 8 \\ 6 & 6 & 6 & 6 \\ 3 & 7 & 3 & 7 \end{pmatrix} \text{ and } c = \begin{pmatrix} 0.1 \\ 0.2 \\ 0.2 \\ 0.4 \\ 0.4 \end{pmatrix}$$

the problem is

$$\min - \sum_{i=1}^{5} \left( c_i + \sum_{j=1}^{4} (x_j - A_{ij})^2 \right)^{-1}$$

$$s.t. \quad \begin{pmatrix} x_1 \\ x_2 \\ x_3 \\ x_4 \end{pmatrix} \in \left[ \begin{pmatrix} 0 \\ 0 \\ 0 \\ 0 \end{pmatrix}, \begin{pmatrix} 10 \\ 10 \\ 10 \\ 10 \end{pmatrix} \right].$$

| $x^L$ | [ 0; 0; 0; 0] |
|---|---|
| $x^U$ | [10;10;10;10] |
| fglob | -10.1531996790582 |
| nglob | 1 |
| xglob | [4; 4; 4; 4] |

**Test problem "Hartman 3"**[9], abbr: **hm3**, Dimension: **3**

With the parameters

$$A = \begin{pmatrix} 3.0 & 10 & 30 \\ 0.1 & 10 & 35 \\ 3.0 & 10 & 30 \\ 0.1 & 10 & 35 \end{pmatrix}, P = \begin{pmatrix} 0.3689 & 0.1170 & 0.2673 \\ 0.4699 & 0.4387 & 0.7470 \\ 0.1091 & 0.8732 & 0.5547 \\ 0.0381 & 0.5743 & 0.8828 \end{pmatrix} \text{ and } c = \begin{pmatrix} 0.1 \\ 1.2 \\ 3.0 \\ 3.2 \end{pmatrix}$$

the problem is

$$\min - \sum_{i=1}^{4} c_i \exp \left( - \sum_{j=1}^{3} A_{ij} (x_j - P_{ij})^2 \right)$$

$$s.t. \quad \begin{pmatrix} x_1 \\ x_2 \\ x_3 \end{pmatrix} \in \left[ \begin{pmatrix} 0 \\ 0 \\ 0 \end{pmatrix}, \begin{pmatrix} 1 \\ 1 \\ 1 \end{pmatrix} \right].$$

| $x^L$ | [0; 0; 0] |
|---|---|
| $x^U$ | [1; 1; 1] |
| fglob | -3.86278214782076 |
| nglob | 1 |
| xglob | [0.1; 0.55592003; 0.85218259] |

**Test problem "Hartman 6"** [9], abbr: **hm6**, Dimension: **6**

With the parameters

$$A = \begin{pmatrix} 10 & 3.0 & 17 & 3.5 & 1.7 & 8 \\ 0.05 & 10 & 17 & 0.1 & 8 & 14 \\ 3.0 & 3.5 & 1.7 & 10 & 17 & 8 \\ 17 & 8 & 0.05 & 10 & 0.1 & 14 \end{pmatrix},$$

$$P = \begin{pmatrix} 0.1312 & 0.1696 & 0.5569 & 0.0124 & 0.8283 & 0.5886 \\ 0.2329 & 0.4135 & 0.8307 & 0.3736 & 0.1004 & 0.9991 \\ 0.2348 & 0.1451 & 0.3522 & 0.2883 & 0.3047 & 0.6650 \\ 0.4047 & 0.8828 & 0.8732 & 0.5743 & 0.1091 & 0.0381 \end{pmatrix} \text{ and } c = \begin{pmatrix} 1.0 \\ 1.2 \\ 3.0 \\ 3.2 \end{pmatrix}$$

the problem is

$$\min \ -\sum_{i=1}^{4} c_i \exp\left( -\sum_{j=1}^{6} A_{ij}\,(x_j - P_{ij})^2 \right)$$

$$s.t. \quad \begin{pmatrix} x_1 \\ x_2 \\ x_3 \\ x_4 \\ x_5 \\ x_6 \end{pmatrix} \in \left[ \begin{pmatrix} 0 \\ 0 \\ 0 \\ 0 \\ 0 \\ 0 \end{pmatrix}, \begin{pmatrix} 1 \\ 1 \\ 1 \\ 1 \\ 1 \\ 1 \end{pmatrix} \right].$$

| | |
|---|---|
| $x^L$ | [0; 0; 0; 0; 0; 0] |
| $x^U$ | [1; 1; 1; 1; 1; 1] |
| fglob | -3.32236801141551 |
| nglob | 1 |
| xglob | [ 0.20168952 |
| | 0.15001069 |
| | 0.47687398 |
| | 0.27533243 |
| | 0.31165162 |
| | 0.65730054 ] |

# Bibliography

[1] C.S. Adjiman, S. Dallwig, C.A. Floudas, and A. Neumaier. A global optimization method, αBB, for general twice-differentiable constrained NLPs - I. theoretical advances. *Computers and Chemical Engineering*, 22(9):1137–1158, 1998.

[2] I.G. Akrotirianakis and C.A. Floudas. A new class of improved convex underestimators for twice continuously differentiable constrained NLPs. *Journal of Global Optimization*, 30(4):367–390, 2004.

[3] I.G. Akrotirianakis and C.A. Floudas. Computational experience with a new class of convex underestimators: box-constrained NLP problems. *Journal of Global Optimization*, 29(3):249–264, 2004.

[4] I.P. Androulakis, C.D. Maranas, and C.A. Floudas. αBB: a global optimization method for general constrained nonconvex problems. *Journal of Global Optimization*, 7(4):337–363, 1995.

[5] D. Banholzer, J. Fliege, and R. Werner. Enhanced calibration of the Nelson-Siegel and the Svensson model. Preprint 2015.

[6] M. Björkman and K. Holmström. Global optimization of costly nonconvex functions using radial basis functions. *Optimization and Engineering*, 1(4):373–397, 2000.

[7] A.J. Booker, J.E.Jr. Dennis, P.D. Frank, D.B. Serafini, and M.W. Trosset. A rigorous framework for optimization of expensive functions by surrogates. *Structural and Multidisciplinary Optimization*, 17(1):1–13, 1999.

[8] P.J.C. Dickinson. On the exhaustivity of simplicial partitioning. *Journal of Global Optimization*, 58(1):189–203, 2014.

[9] L.C.W. Dixon and G.P. Szegö. *Towards global Optimisation 2*. North-Holland, Amsterdam, 1978.

[10] S.D. Flåm, H.Th. Jongen, and O. Stein. Slopes of shadow prices and Lagrange multipliers. *Optimization Letters*, 2:143–155, 2008.

[11] C.A. Floudas and C.E. Gounaris. A review of recent advances in global optimization. *Journal of Global Optimization*, 45(1):3–38, 2009.

[12] C.E. Gounaris and C.A. Floudas. Tight convex underestimators for $\mathcal{C}^2$-continuous problems: I. univariate functions. *Journal of Global Optimization*, 42:51–67, 2008.

[13] H.-M. Gutmann. A radial basis function method for global optimization. *Journal of Global Optimization*, 19(3):201–227, 2001.

[14] H.-M. Gutmann. *Radial basis function methods for global optimization*. PhD thesis, University of Cambridge, 2001.

[15] F.J. Hall and C.D. Meyer. Generalized inverses of the fundamental bordered matrix used in linear estimation. *Sankhya: The Indian Journal of Statistics, Series A*, 37(3):428–438, 1975.

[16] P. Hansen, B. Jaumard, and S.H. Lu. On using estimates of Lipschitz constants in global opimization. *Journal of Optimization Theory and Applications*, 75(1):195–200, 1992.

[17] R.L. Hardy. Multiquadric equations of topography and other irregular surfaces. *Journal of Geophysical Research*, 76:1905–1915, 1971.

[18] K. Holmström. An adaptive radial basis algorithm (ARBF) for expensive black-box global optimization. *Journal of Global Optimization*, 41(3):447–464, 2008.

[19] R. Horst. On generalized bisection of $n$-simplices. *Mathematics of Computation*, 66(218):691–698, 1997.

[20] R. Horst, P.M. Pardalos, and N.V. Thoai. *Introduction to global optimization*. Kluwer Academic Publishers, Dordrecht, 2nd edition, 2000.

[21] R. Horst and H. Tuy. *Global optimization, deterministic approaches*. Springer-Verlag Berlin Heidelberg, 1996.

[22] S. Jakobsson, M. Patriksson, J. Rudholm, and A. Wojciechowski. A method for simulation based optimization using radial basis functions. *Optimization and Engineering*, 11(4):501–532, 2009.

[23] M. James. The generalised inverse. *Mathematical Gazette*, 62:109–114, 1978.

[24] D.R. Jones. A taxonomy of global optimization methods. *Journal of Global Optimization*, 21(4):345–383, 2001.

[25] D.R. Jones, C.D. Perttunen, and B.E. Stuckman. Lipschitzian optimization without the Lipschitz constant. *Journal of Optimization Theory and Application*, 79(1):157–181, 1993.

[26] D.R. Jones, M. Schonlau, and W.J. Welch. Efficient global optimization of expensive black-box functions. *Journal of Global Optimization*, 13(4):455–492, 1998.

[27] J.H. Kämpf, M. Wetter, and D. Robinson. A comparison of global optimisation algorithms with standard benchmark functions and real-world applications using EnergyPlus. *Journal of Building Performance Simulation*, 3(2):103–120, 2010.

[28] H.J. Kushner. A new method of locating the maximum point of an arbitrary multipeak curve in the presence of noise. *Journal of Basic Engineering*, 86(1):97–106, 1964.

[29] L. Liberti and C.C. Pantelides. Convex envelopes of monomials of odd degree. *Journal of Global Optimization*, 25:157–168, 2003.

[30] M.D. McKay, R.J. Beckman, and W.J. Conover. A comparison of three methods for selecting values of input variables in the analysis of output from a computer code. *Technometrics*, 21(2):239–245, 1979.

[31] C.C. Meewella and D.Q. Mayne. An algorithm for global optimization of Lipschitz continuous functions. *Journal of Optimization Theory and Applications*, 57(2):307–322, 1988.

[32] C.A. Meyer and C.A. Floudas. Convex underestimation of twice continuously differentiable functions by piecewise quadratic pertubation: spline $\alpha$BB underestimators. *Journal of Global Optimization*, 32(2):221–258, 2005.

[33] S.K. Mitra. *Properties of the fundamental bordered matrix used in linear estimation, In: Statistics and Probability, Essays in Honor of C.R. Rao (G. Kallianpur et al. eds.)*. North-Holland, 1982.

[34] B. Noble and J.W. Daniel. *Applied linear algebra*. 3rd edition, Prentice-hall / Englewood Cliffs, NJ07632, 1988.

[35] M.J.D. Powell. *Approximation theory and methods*. Cambridge University Press, 1981.

[36] M.J.D. Powell. The theory of radial basis function approximation in 1990. *In: W.A. Light (ed.), Advances in Numerical Analysis*, 2: Wavelets, Subdivision Algorithms, and Radial Basis Functions:105–210, 1992.

[37] M.J.D. Powell. Recent research at Cambridge on radial basis functions. *International Series of Numerical Mathematics*, 132:215–232, 1999.

[38] R.G. Regis and C.A. Shoemaker. Constrained global optimization of expensive black box functions using radial basis functions. *Journal of Global Optimization*, 31(1):153–171, 2005.

[39] R.G. Regis and C.A. Shoemaker. Improved strategies for radial basis function methods for global optimization. *Journal of Global Optimization*, 37(1):113–135, 2007.

[40] R. Schaback. Radial basis functions viewed from cubic splines. *In: Multivariate Approximation and Splines, G. Nürnberger, J.W. Schmidt and G.Walz (eds), Birkhäuser Verlag, Basel*, pages 245–258, 1997.

[41] R. Schaback. Native Hilbert spaces for radial basis functions I. *International Series of Numerical Mathematics*, 132:255–282, 1999.

[42] H.D. Sherali and V. Ganesan. A pseudo-global optimization approach with application to the design of containerships. *Journal of Global Optimization*, 26:335–360, 2003.

[43] C. Stanisch. Globale Optimierung bei teuren Funktionsauswertungen - ein Verfahren mit Hilfe radialer Basisfunktionen, Diplomarbeit, Universität Dortmund, 2006.

[44] A. Žilinskas. Axiomatic approach to statistical models and their use in multimodal optimization theory. *Mathematical Programming*, 22:104–116, 1982.

[45] H. Wendland. *Scattered data approximation*. Cambridge University Press: Cambridge Monographs on Applied and Computational Mathematics, 2005.

[46] S.M. Wild, R.G. Regis, and C.A. Shoemaker. ORBIT: Optimization by radial basis function interpolation in trust-regions. *SIAM Journal on Scientific Computing*, 30(6):3197–3219, 2008.

[47] Z.B. Zabinsky. *Stochastic adaptive search for global optimization*. Kluwer Academic Publishers, Boston Dordrecht London, 2003.

# Acknowledgements

Writing this PhD was influenced by many accidental or planned moments in my life and of course by the people who supported me directly and indirectly. I take this opportunity to express my gratitude to some of them.

First of all I would like to thank my supervisor, Prof. Dr. Mirjam Dür, whose excellent and continuous mentoring enabled me to improve and unfold my ability in doing research work. I would also like to thank Prof. Dr. Ralf Werner for reading my thesis and giving helpful comments.

When I look back, I remember a few special episodes and moments. These are uniquely connected to people who I am grateful to have met and who, maybe without knowing, directed my way at some stage. I would like to thank Reinhard Koll for facilitating my interest in mathematics at school, Teuta Ilazi for her help to find the right field of profession and Sebastian Stranz for his encouraging words during my time in Hagen.

At my time at the University of Trier I met a lot of people, who made my time in Trier enjoyable. Some of these people are my former colleagues at the Department of Mathematics. I would like to thank especially Van Nguyen and Ulf Friedrich for inspiring talks, Monika Thieme-Trapp for her help with administrative matters, as well as the members of the SIAM Student Chapter, for a continual pleasant working ambience.

I am very grateful for the stimulating help of my friends, with a special thanks to Antje Pack for spending a lot of hours proofreading my English grammar. I would also like to express my gratitude to my parents and my sister, who taught me not to take things for granted.

Finally I would like to thank my family for a warm atmosphere which enabled me to write this thesis: I thank my daughters Nellie and Nora, who did – and still do – make my life so much more colorful. I deeply thank Sebastian for his unconditional and reliable support in the process of writing my PhD.

*Åkersberga, August 2016*                                                    Christine Edman